AF842637

No Code:No Problem

Build Your Website & Chatbot with WordPress & Botpress

by
Susie Hala

Independently Published

🕮 *About the Author*

Susie Hala is an entrepreneur, educator, and author dedicated to making technology simple and accessible for learners at every level. She is the founder of **YourAIStudyBuddy.com**, an online platform that provides step-by-step resources to help students, professionals, and communities understand artificial intelligence, digital tools, and emerging technologies.

Susie is the author of *IntelliGloss: The Definitive AI Dictionary*, a comprehensive 1,000+ page reference book that has reached readers across more than 20 countries, and *Raquel and Elfie Love: A Tale of Strength*, her first children's book published worldwide. With a background in business, publishing, and education, she brings her passion for technology and her gift for clear explanations together to empower non-technical audiences.

Through her work, Susie continues to blend her Pacific Island heritage, entrepreneurial spirit, and dedication to lifelong learning. Whether writing for students, professionals, or entire communities, her mission is to bridge the gap between complex technology and everyday understanding.

✍ *Note from the Author*

When I first set out to build websites and chatbots, I wasn't a "tech person." I was just someone with ideas, determination, and the willingness to learn step by step. Along the way, I realized how confusing technical jargon, hidden settings, and complex explanations could be for beginners. That experience inspired me to write this book.

No Code, No Problem is for the dreamers, entrepreneurs, educators, and students who want to create something online without feeling left behind by technology. My goal is simple: to give you a clear roadmap, written in plain language, so you can build with confidence.

Thank you for trusting me to be your guide on this journey. I hope these pages give you not only the tools but also the encouragement to take your ideas from imagination to reality.

— **Susie Hala**

📖 *Preface*

Technology is shaping every corner of our lives, but for many people, the journey to building a website or chatbot feels overwhelming. Too often, the conversation is dominated by technical experts, leaving non-technical creators feeling like they don't belong. I know this feeling well — because I once stood in that exact place.

This book was born out of my own struggles and discoveries while building online platforms without a background in coding. Step by step, I learned that tools like WordPress, Hostinger, Stripe, and Botpress can empower anyone — even those with no technical training — to create professional websites and intelligent chatbots. Along the way, I also realized that what people needed most was not just tutorials, but *clarity*.

No Code, No Problem is written with you in mind: the entrepreneur who wants to sell online, the educator who wants to share knowledge, the dreamer who wants to bring an idea to life. My goal is not only to teach you which buttons to click but to give you the confidence that you belong in the digital space.

As you read, remember this: you don't need to know everything to begin. Start small, follow the steps, and keep moving forward. By the end of this book, you'll have not only a website and chatbot, but also the confidence to keep building, experimenting, and learning.

—

Susie Hala

For permissions or inquiries, contact:
📧 info@youraistudybuddy.com
📍 Published in the United States of America

This book is independently published by Susie Hala through IngramSpark.
ISBN: 979-8-9991963-1-6

Trademark Acknowledgments

1. WordPress® is a registered trademark of the WordPress Foundation.
2. Stripe® is a registered trademark of Stripe, Inc.
3. Botpress® is a registered trademark of Botpress, Inc.
4. Hostinger® is a registered trademark of Hostinger International Ltd.
5. Google® is a registered trademark of Google LLC.

This publication is not affiliated with, endorsed by, or sponsored by WordPress, Automattic, Stripe, Botpress, Hostinger, Google, or any other company mentioned. All product names, logos, and brands are property of their respective owners. Any references to specific software platforms or services are made strictly for educational and illustrative purposes only.

Disclaimer

The content in this book is provided for educational and informational purposes only. The author has made every effort to ensure the accuracy of information at the time of publication. However, technologies evolve, and platform features such as those offered by WordPress, Stripe, Botpress, and Hostinger may change over time.

The author assumes no responsibility or liability for any errors, omissions, or results from the use of this content. This book does not offer legal, financial, or professional advice, and readers are encouraged to consult with qualified professionals when making technical, business, or financial decisions related to website development, e-commerce, or AI integratigrating.

Acknowledgements

This book was made possible by the encouragement and support of those who believe that learning never stops—no matter your age, background, or technical skill.

To my grandchildren, who are my inspiration and motivation. Your questions and love for discovery keep me growing.

To my friends and family—thank you for your patience, support, and love during the late nights and early mornings it took to finish this.

To the global WordPress community—thank you for building tools that empower non-technical people like me to dream, design, and publish something meaningful.

Table of Contents

A Note to My Readers: Why Some Topics Are Repeated

As you read through this book, you may notice that certain concepts—like connecting your WordPress site to Hostinger, integrating Stripe for payments, adding your Botpress chatbot, or optimizing your site with SEO and Google tools—appear in more than one chapter.

This is intentional: In the real world, these tools don't operate in isolation. They work together, and often, setting up one feature depends on another. For example:

1. You can't connect Stripe without first setting up your website hosting.
2. You can't get found on Google without connecting your domain and submitting your sitemap.
3. Your chatbot can't help customers if your pages aren't organized or your SEO isn't working.

Because of this overlap, I've chosen to repeat key steps or explanations in different sections. I do this not to fill pages, but to make sure that each chapter can stand on its own—especially for readers who might jump to the section they need most.

This repetition is here to:

1. Reinforce what's important
2. Help you gain confidence
3. Reduce confusion as you build your own website or online business
4. Whether you're a beginner or a business owner learning as you go, I want this book to feel clear, supportive, and easy to follow—even if it means saying things more than once.Thank you for trusting me to walk with you step by step. Let's build something amazing together.

Warmly,

Susie Hala

🌱 The Family Section

Before diving into the world of artificial intelligence, it's important to meet the **families of technology** that make everything possible. Just like a human family has parents, children, and generations that grow larger, the digital world is built on its own "family tree." These families give us the structure to understand how computers think, how they store information, and how they power AI.

The Bit Family and How Computers Work Behind the Scenes

Introduction

Every website—including your WordPress site—runs on the smallest digital ingredient: the **bit**. Just like flour is the foundation of bread, bits (0s and 1s) are the foundation of computing. Every photo you upload, every WordPress post you publish, and every transaction on your site is built from billions of these bits.

To understand how WordPress works, let's follow the journey of a bit: from programming languages, to the CPU, to memory, to hosting companies where your site lives, and finally to your visitors' screens.

The Bit and Its Family

- **Bit (b):** The smallest unit, 0 or 1.
- **Byte (B):** 8 bits, often enough to store one character.
- **Kilobyte (KB):** About a paragraph of text.
- **Megabyte (MB):** A photo or a few minutes of audio.
- **Gigabyte (GB):** A movie or a full WordPress backup.
- **Terabyte (TB):** Entire websites or large hosting plans.
- Beyond: **Petabytes, Exabytes, Zettabytes**—used in cloud storage and big data.

These are the building blocks of the digital world. Hosting companies rent out space measured in gigabytes or terabytes so your WordPress files and databases (billions of bits) have a home.

How Computers Process Bits

1. **Programming Languages → Binary**
   Humans write code in PHP, JavaScript, or Python. But computers only understand binary. Translators (compilers or interpreters) convert programming instructions into 0s and 1s.

👉 Example: A WordPress plugin written in PHP is transformed into binary bits the CPU can execute.

2. **CPU: The Chef of Bits**
   The CPU (Central Processing Unit) is like a chef in a kitchen. It takes binary instructions (recipes) and processes them step by step. Every blog post save, login request, or WordPress calculation is handled by the CPU one bit at a time.

3. **Memory and Storage**

RAM (temporary storage): Like a countertop, where bits are processed while tasks are active. If too many visitors open your WordPress site, more RAM is needed to handle all those bits.

Hard Drive / SSD (permanent storage): Like a pantry, where bits are stored long-term. Your WordPress themes, plugins, and database live here until needed.

The Hosting Company's Role

When you sign up for a hosting plan (e.g., Hostinger, Bluehost, SiteGround), you are renting space on their powerful computers—called **servers**. Here's what happens behind the scenes:

- **File Storage:** Your WordPress files (themes, plugins, uploads) are stored as billions of bits on the hosting company's hard drives or SSDs.
- **Database Storage:** Your posts, comments, and settings are stored as structured bits in databases (like MySQL).
- **Processing Power:** The hosting company's CPUs execute the WordPress code every time someone visits your site.
- **Memory Allocation:** Hosting assigns RAM to keep active processes (visitors browsing, checkout pages, etc.) running smoothly.
- **Network & Delivery:** Hosting servers are connected to the internet backbone. When someone types your domain, the hosting server sends the correct bits across the web to display your page.

- **Backups & Security:** Hosting companies also keep copies (backups) of your bits and protect them with firewalls to keep hackers from stealing or corrupting them.

👉 Example: When a visitor loads your homepage, the hosting company's server fetches the bits for your WordPress theme, your latest blog post, and your images, and then sends those bits back to the visitor's browser.

Input and Output

Input: Your keyboard strokes (like typing a blog post) are turned into bits and sent to the server.

- **Output:** Hosting servers process those bits and send them back as HTML, CSS, and JavaScript that your visitor's browser can display.

WordPress in Action (Full Cycle)

1. You type a blog post (input → bits).
2. CPU processes it (bits executed).
3. Hosting server stores it (bits saved in database).
4. A visitor requests the page → server retrieves the bits.
5. Browser translates bits into text, images, and layout (output).

Why It Matters for Website Owners

- **Storage Plans = Bit Limits:** Hosting plans (1 GB, 10 GB, 100 GB) limit how many bits your site can store.
- **CPU Power = Site Speed:** Faster CPUs handle more bits per second, making your WordPress site load faster.
- **RAM Allocation = Handling Visitors:** More memory means more people can browse your site at the same time without crashing.
- **Data Centers = Global Reach:** Hosting companies place servers around the world to deliver bits faster to users, no matter where they are.

In short, hosting companies are like the landlords of the digital world. You rent space on their servers, and they make sure your bits are safe, accessible, and delivered quickly to your visitors.

The CPU Family:

How the Computer's Brain Works

Introduction

If the **bit** is the smallest ingredient of computing, then the **CPU (Central Processing Unit)** is the master chef that brings everything together. Often called the "brain" of the computer, the CPU is responsible for interpreting instructions, carrying out calculations, and managing the flow of data. Without the CPU, bits and bytes would just sit there, like ingredients on a shelf with no one to cook them.

Every action on your WordPress site—logging in, uploading an image, processing a checkout, or publishing a blog post—is powered by the CPU's ability to process information at lightning speed. Understanding how it works helps you appreciate why hosting power, RAM, and optimization matter so much for your website.

How the CPU Processes Information

1. **Fetch**
   - The CPU starts by **fetching instructions** from memory (RAM).
   - These instructions might be part of your WordPress code, such as "Load the homepage layout."

   2. Decode

   - Once fetched, the CPU **decodes the instruction** into binary (0s and 1s).
   - This translation is like breaking down a recipe into small, actionable cooking steps.
2. **Execute**
   - The CPU **executes the instruction**.
   - This could mean doing math (adding two numbers), making a decision (if this, then that), or moving data around (save a post to the database).
3. **Store (or Output)**
   - After execution, the CPU **stores the result** in memory or sends it to an output device.
   - Example: When you click *Publish* in WordPress, the CPU processes that command, stores the blog post in the database, and then sends the confirmation message back to your screen.

Key Components Inside the CPU

- **ALU (Arithmetic Logic Unit):** Handles math and logic, such as comparing values or running calculations.

- **CU (Control Unit):** The traffic director—it tells other components what to do and when.
- **Registers:** Tiny, ultra-fast storage areas inside the CPU for holding instructions and data mid-process.
- **Cache:** High-speed memory close to the CPU, storing frequently used data for quick access

Everyday Analogy

Think of the CPU as the **chef in a busy kitchen**:

- **Recipe Book = Code (PHP, JavaScript, etc.)**
- **Ingredients = Bits (0s and 1s)**
- **Chef = CPU**
- **Countertop = RAM (workspace for active tasks)**
- **Pantry = Hard Drive (permanent storage)**
- **Served Dish = Output on your screen**

The chef doesn't cook the whole menu at once—he fetches one recipe, breaks it into steps, cooks it, and serves it. The CPU does the same, except it repeats this billions of times per second.

Why This Matters for WordPress and Hosting

When choosing hosting for WordPress, the **CPU power** of the server directly affects your site's performance:

- A stronger CPU processes more bits at once, making your site faster.
- CPUs with larger caches handle complex tasks (like WooCommerce checkouts) more smoothly.
- Shared hosting means your site's CPU resources are limited, while VPS or dedicated hosting gives your site more CPU access.

In short, the CPU is the engine under the hood. The stronger it is, the faster and smoother your WordPress site runs

The Bit Family

At the very bottom of the digital world lives the **bit**, the smallest unit of information. A bit can only be a **0** or a **1** — like a light switch that's either off or on. By itself, a bit doesn't say much, but when grouped together, bits form something bigger. Eight bits make one **byte**, which can represent a single letter, number, or symbol. From there, the Bit Family keeps growing: kilobytes, megabytes, gigabytes, terabytes, and beyond. Each generation is one thousand times larger than the one before, until we reach **yottabytes**, which are so massive they can hold centuries of human knowledge. The Bit Family shows us that the digital world grows step by step, just like a family tree.

The Bit Family: From Tiny 0s and 1s to the Fuel of AI

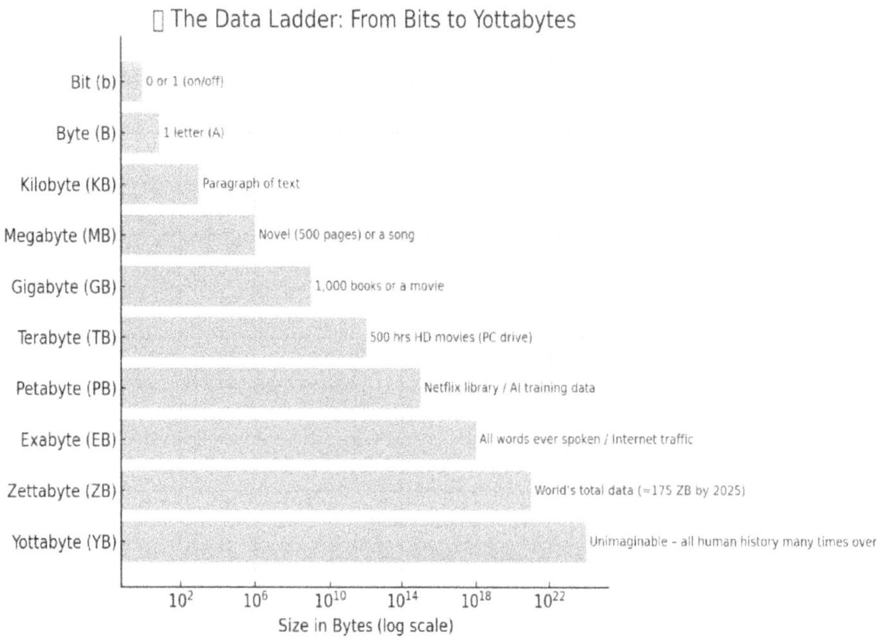

What Is a Bit?

A **bit** is the smallest piece of information a computer can understand.

- It can only be a **0** or a **1**.
- Think of it like a light switch: **off (0)** or **on (1)**.
- Alone, a bit isn't very powerful — but together, bits build everything in the digital world.

2. From Bit to Byte

When you group **8 bits together**, you get a **byte**.

- A byte can represent a single **letter** (like "A") or a number.
- Example:
  - `01000001` in binary = the letter **A**.

Bytes are the "words" computers use to talk.

3. The Growing Bit Family

Here's how the family grows, each step getting 1,000 times bigger:

- **1 Byte (B) = 8 bits** → one letter.
- **1 Kilobyte (KB) = 1,000 bytes** → a short page of text.
- **1 Megabyte (MB) = 1,000 KB** → a whole book or a song.
- **1 Gigabyte (GB) = 1,000 MB** → a movie or 1,000 books.
- **1 Terabyte (TB) = 1,000 GB** → 500 hours of HD movies.
- **1 Petabyte (PB) = 1,000 TB** → Netflix's movie library.
- **1 Exabyte (EB) = 1,000 PB** → all words ever spoken by humans.
- **1 Zettabyte (ZB) = 1,000 EB** → almost all the world's internet data.
- **1 Yottabyte (YB) = 1,000 ZB** → beyond imagination — centuries of all human knowledge

4. Why the Bit Family Matters for AI

AI is **hungry for data**.

- Training an AI like ChatGPT requires **petabytes** of text, images, and videos.
- The larger the AI, the more **bits and bytes** it consumes.
- Without the Bit Family, there is no way to measure or manage this massive scale of information.

Examples in AI:

- **Image recognition AI**: A single high-resolution picture can be several MBs. Millions of images = TBs.
- **Language models**: Trained on trillions of words = PBs of text.
- **Video AI (self-driving cars)**: Endless hours of footage = PBs to EBs.

5. Bits and Compute Power

It's not only about storage. AI chips (like GPUs and TPUs) also process bits at incredible speed:

- Modern GPUs can handle **trillions of bits per second**.
- The more efficiently a chip moves bits, the faster an AI can learn.

This is why companies like **NVIDIA** and **Google** invest in specialized chips. AI depends on both **how much data (bits/bytes)** and **how fast chips can crunch them**.

6. The Bigger Picture

The Bit Family is like the **currency of the AI world**:

- **Bits** are the smallest coins.
- **Bytes, KBs, MBs** are the dollars and hundreds.
- **GBs, TBs, PBs** are the millions.
- **EBs, ZBs, YBs** are the billions and trillions.

Without this family, we couldn't measure, store, or train AI at all.

💡 Key Takeaway for Readers

The Bit Family is **not just numbers** — it is the foundation of everything digital. From your smartphone storing a photo to an AI model training on petabytes of data, it all comes down to **0s and 1s growing into giants**.

Understanding this family helps you see why AI requires so much **data, compute, and energy** — and why leaders like Jensen Huang (NVIDIA) and Elon Musk keep talking about **petabytes and exabytes**.

💾 The Data Family

The **Data Family** tells us how much information is being stored or moved. It starts with the smallest unit, the **bit**, and grows step by step into massive amounts like **yottabytes**. Each "generation" of this family is about **1,000 times bigger** than the one before.

☐ Breakdown of the Data Family

- **1 Bit (b):**
  The smallest piece of data — a 0 or 1 (off/on).
- **1 Byte (B) = 8 bits**
  Enough to store a single letter, like "A".
- **1 Kilobyte (KB) = 1,000 bytes**
  A short page of text or a tiny image.
- **1 Megabyte (MB) = 1,000 KB**
  About the size of one song, or a few pages of images.
- **1 Gigabyte (GB) = 1,000 MB**
  A full movie, or thousands of books.

- **1 Terabyte (TB) = 1,000 GB**
  Around 500 hours of HD video, or a personal hard drive.
- **1 Petabyte (PB) = 1,000 TB**
  A data center's worth of storage, or Netflix's movie library.
- **1 Exabyte (EB) = 1,000 PB**
  Roughly all the words ever spoken by humans.

1 Zettabyte (ZB) = 1,000 EB
Almost all the data on the global internet in a year.

- **1 Yottabyte (YB) = 1,000 ZB**
  Beyond imagination — more than all human digital data ever created so far.

💡 Why the Data Family Matters

The Data Family is the **measurement system of the digital world**. Just as humans measure distance with meters or miles, computers measure information with bytes.

In AI, this family is crucial because training models requires **gigabytes to petabytes of data**. For example:

- A small AI model might be trained on gigabytes of text.
- A medium one could need terabytes.
- Giant models (like ChatGPT) are trained on petabytes or more.

The Data Family helps us understand why companies keep building bigger chips, larger memory systems, and massive data centers: AI's hunger for data keeps growing.

From Short-Term to Long-Term Memory in Computers

When you use a computer, it feels fast, smooth, and ready for anything. That's because your device uses different kinds of memory — some designed for **speed in the moment (short-term)**, and some designed for **keeping your data safe for years (long-term).**

Understanding the difference helps you know why adding more RAM makes your computer feel faster, or why upgrading to an SSD makes your files open instantly.

Here's how computer memory works, from short-term to long-term:

1. Cache Memory → Ultra Short-Term Memory

- The fastest memory in a computer.

- Stores tiny bits of data the CPU needs right now.
- Example: speeds up repeat calculations so your processor doesn't have to "rethink" every step.

2. RAM (Random Access Memory) → Short-Term Memory

- The workspace for all your active apps and files.
- Super fast but temporary — it's erased when the power turns off.
- Example: keeping your browser tabs, videos, or documents open while you work.

3. Hard Disk Drives (HDDs) → Traditional Long-Term Storage

- Stores data magnetically on spinning disks.
- Slower than newer storage types, but affordable and holds lots of files.
- Example: keeping old photos, music, or large archives.

4. Solid State Drives (SSDs) → Modern Long-Term Storage

- Uses flash memory (no moving parts).
- Much faster and more reliable than HDDs.
- Example: your laptop booting in seconds or apps opening instantly.

5. Cloud Storage → Long-Term, Accessible Anywhere

- Files are stored on servers across the internet.
- Accessible from any device, anywhere in the world.
- Example: saving a document to Google Drive or iCloud so you can open it on your phone later.

🌸 Plain English Takeaway

- **Cache + RAM = short-term memory** → speed for what you're doing right now.
- **HDD + SSD + Cloud = long-term memory** → safety and access for what you want to keep.

Just like the human brain, computers need *both kinds* of memory: quick recall for the present and reliable storage for the future.

The Chip Family

Chips are the hidden engines of the digital world. They sit inside our phones, laptops, cars, and even in massive data centers, powering everything from text messages to artificial intelligence. In simple terms, a chip is a tiny piece of silicon packed with billions of transistors that work together to process information at incredible speeds.

In the world of AI, chips are more than just hardware — they are the foundation that determines how fast and efficiently machines can learn, think, and respond. From **CPUs** (the all-purpose planners) to **GPUs** (the parallel problem-solvers), and from custom AI processors to experimental wafer-scale designs, each chip plays a unique role in the larger "family" of computing.

It's important to remember that **chips change all the time**. New designs are released every few years — sometimes every few months — as companies race to create faster, smaller, and more energy-efficient processors. A chip that leads the industry today may be replaced tomorrow by a breakthrough with more memory, better energy use, or specialized AI features.

By understanding the Chip Family, you're not just learning about hardware — you're stepping into the heartbeat of modern technology. And while the names and models will continue to evolve, the central idea remains: chips are the essential building blocks of our digital and AI-powered future.

10 AI Chips Companies

1. NVIDIA A100

A data center chip built for accelerating artificial intelligence and high-performance computing. It delivers strong parallel processing for training massive neural networks.

2. NVIDIA H100

The successor to the A100, optimized for generative AI. It uses advanced memory bandwidth and transformer acceleration to handle large-scale models more efficiently.

3. NVIDIA GH200 Grace Hopper Superchip

A hybrid design that combines a GPU with an ARM-based CPU in one package. It reduces data transfer delays and increases speed for memory-intensive AI tasks.

4. Google TPU v4

A custom chip developed by Google specifically for deep learning. It provides large amounts of processing power for model training and inference inside Google Cloud.

5. AMD MI300X

A high-performance GPU designed to compete in AI workloads, offering large memory capacity and efficient parallel computation.

6. Intel Gaudi 2

Created by Intel's Habana Labs, this AI processor focuses on accelerating training and inference while being cost-efficient for enterprises.

7. Apple Neural Engine

Integrated into Apple devices, this chip performs on-device AI tasks such as image recognition, natural language processing, and real-time translation without relying on cloud processing.

8. Tesla Dojo D1

A custom AI chip developed by Tesla for training self-driving car systems. It is designed to process massive video datasets more quickly than traditional GPUs.

9. AWS Trainium

Amazon Web Services' custom AI chip, built to accelerate training of deep learning models in the cloud at scale and reduce costs compared to standard GPUs.

10. Cerebras Wafer-Scale Engine (WSE-2)

One of the largest chips ever built, spanning an entire silicon wafer. It is designed for extremely large AI workloads, such as natural language models and scientific computing.

Measuring Energy Use of GPUs

Step 1: Start with the GPU's Power Rating

Most modern GPUs use about **300 watts (W)** when running at full power.

- **Watts (W):** A measure of power, or "how fast energy is being used."
- Example: A light bulb = 60 W. A GPU = 300 W.

Step 2: Convert Watts to Kilowatts

Since energy use is often measured in **kilowatt-hours (kWh)** (like on your electricity bill):

- 1 kilowatt (kW) = 1,000 watts.
- **300 W ÷ 1,000 = 0.3 kW.**

So, **1 GPU = 0.3 kW.**

Step 3: Energy Over Time (kWh)

Energy = Power × Time.

- **1 GPU running for 1 hour:**
  0.3 kW × 1 hour = **0.3 kWh**
- **1 GPU running for 10 hours:**
  0.3 kW × 10 hours = **3 kWh**
- **1 GPU running for 24 hours (1 day):**
  0.3 kW × 24 hours = **7.2 kWh**
- **1 GPU running for 30 days (1 month):**
  0.3 kW × 24 hours × 30 = **216 kWh**

Step 4: Scale to Multiple GPUs

Let's see how this looks when you run **more GPUs at once** (all at 300 W each):

Number of GPUs	Total Watts	Kilowatts (kW)	Energy in 1 Hour (kWh)	Energy in 1 Day (kWh)	Energy in 30 Days (kWh)
1 GPU	300 W	0.3 kW	0.3 kWh	7.2 kWh	216 kWh
5 GPUs	1,500 W	1.5 kW	1.5 kWh	36 kWh	1,080 kWh
10 GPUs	3,000 W	3.0 kW	3 kWh	72 kWh	2,160 kWh
20 GPUs	6,000 W	6.0 kW	6 kWh	144 kWh	4,320 kWh

Step 5: Real-World Comparison

- **1 household in the U.S. uses ~30 kWh/day.**
- So:
  - **1 GPU (7.2 kWh/day)** = ¼ of a household's daily use.
  - **10 GPUs (72 kWh/day)** = energy for **2–3 households/day.**

- **20 GPUs (144 kWh/day)** = energy for **about 5 households/day.**

🌸 Plain English

- 1 GPU at 300 watts doesn't sound like much.
- But when you run 10, 20, or 1,000 GPUs **nonstop for weeks**, the energy use explodes — enough to power entire towns.

The GPU Family

GPUs (Graphics Processing Units) are the engines behind video games, visual effects, and now—artificial intelligence. Just like the **Bit Family grows from a single bit to zettabytes**, the GPU Family grows from small chips in laptops to giant clusters powering global AI models. Let's meet the members of the GPU Family.

1. Baby GPU – The Integrated Chip

- Lives inside CPUs like Intel or Apple's M-series.
- Shares memory with the computer, not very powerful on its own.
- Great for browsing, Netflix, or basic graphics.
- **Analogy**: The **toddler of the family** — small, dependent, and learning.

2. Teen GPU – The Gaming Card

- Examples: NVIDIA GeForce, AMD Radeon.
- Found in gaming PCs and laptops.
- Handles complex graphics, VR, and even beginner AI projects.
- **Analogy**: The **teenager** — fast, energetic, and experimenting with hobbing

3. Parent GPU – The Workstation Pro

- Examples: NVIDIA RTX A-series, AMD Radeon Pro.
- Used by architects, engineers, and video editors.
- Built for long work hours and creative heavy-lifting.
- **Analogy**: The **working parent** — responsible, professional, and reliable.

4. Grandparent GPU – The Data Center Giant

- Examples: NVIDIA A100, H100, AMD MI300.

- Lives in server farms and powers AI training (like GPT models).
- Handles trillions of calculations per second.
- **Analogy**: The **wise elder** — strong, powerful, guiding the younger generation.

5. Cousin GPU – The Specialist

- Examples: Google TPU, Graphcore IPU.
- Not exactly a GPU, but designed for AI tasks.
- Runs specific workloads (like training neural networks).
- **Analogy**: The **cousin who chose a unique career** — still family, but specialized.

6. Family Reunion – The Supercomputer Cluster

- Example: NVIDIA DGX systems, supercomputers like Frontier.
- Thousands of GPUs linked together.
- Train massive AI models like GPT-4 or future GPT-5.
- **Analogy**: The **family gathering** — everyone teams up to achieve something extraordinary.

Why the GPU Family Matters

The GPU Family explains why AI is booming:

- Baby GPUs make laptops affordable.
- Teen GPUs fuel gaming and creativity.
- Parent and Grandparent GPUs make AI possible.
- Cousins and reunions push technology to the global stage.

Just like every family member has a role, every GPU has its place — from running Minecraft to training ChatGPT.

The "Ingredients" for Superintelligence

Superintelligence isn't here yet, but experts believe it would require several key ingredients:

1. Massive Neural Networks (the brain power)

- Much larger and more advanced than today's AI.
- Networks that can learn not just one task, but many — and connect them together.

2. General Intelligence (flexibility)

- The ability to apply knowledge across different fields, like humans do.

- Example: learning chess *and* then using that reasoning for business or science.

3. Memory and Knowledge (long-term learning)

- AI that remembers past experiences and improves over time.
- A kind of "lifelong learning," not just single-task training.

4. Creativity and Imagination

- Beyond logic: the ability to invent new ideas, art, or science.
- Not just solving problems, but asking new questions.

5. Ethics and Alignment (the compass)

- Without moral direction, superintelligence could be dangerous.
- It needs values, goals, and safety systems aligned with humanity.

6. Energy and Hardware (the fuel)

- Superintelligence would need enormous computing power and efficient chips.
- Future versions of GPUs, TPUs, quantum computers, or new architectures might be required.

🏵 Plain English

The ingredients for superintelligence are: **a bigger brain (advanced neural networks), flexible thinking (general intelligence), memory, creativity, ethics, and powerful hardware.**

The CPU Family

The CPU (Central Processing Unit) is often called the *brain of the computer*. Just like families grow from toddlers to grandparents, the CPU Family ranges from tiny chips in smart devices to powerhouse processors running global systems. Let's meet them.

1. Baby CPU – The Microcontroller

- Lives inside small devices: microwaves, smart bulbs, remote controls.
- Handles one job at a time (like turning on lights).
- **Analogy**: The **infant of the family** — small, simple, focused on basic tasks.

2. Teen CPU – The Laptop & Desktop Processor

- Examples: Intel Core i5, Apple M1, AMD Ryzen 5.
- Runs everyday tasks like web browsing, Office, video streaming.
- Can even multitask with light coding or gaming.
- **Analogy**: The **teenager** — fast, curious, and balancing school (basic tasks) with hobbies (creative work).

3. Parent CPU – The Workstation Powerhouse

- Examples: Intel Core i9, AMD Ryzen 9, Apple M2 Max.
- Built for professionals: video editors, engineers, developers.
- Handles multitasking, heavy applications, and simulations.
- **Analogy**: The **busy parent** — working hard, keeping the household (your computer) running.

4. Grandparent CPU – The Server & Data Center Processor

- Examples: Intel Xeon, AMD EPYC.
- Runs cloud platforms, websites, and enterprise apps.
- Supports hundreds or thousands of users simultaneously.
- **Analogy**: The **grandparent** — experienced, stable, supporting the whole extended family (global businesses).

5. Cousin CPU – The Mobile & Embedded Processor

- Examples: Qualcomm Snapdragon, Apple A-series, Samsung Exynos.
- Found in smartphones, tablets, and cars.
- Optimized for low power and efficiency, but still powerful.
- **Analogy**: The **cousin who's always traveling** — portable, flexible, always connected.

6. Family Reunion – The Supercomputer CPU Cluster

- Example: Fujitsu A64FX in Japan's Fugaku supercomputer.
- Tens of thousands of CPUs working together on scientific research.
- Solve massive problems: climate models, medical simulations, AI training.
- **Analogy**: The **family gathering** — every member joins forces to tackle challenges too big for one alone.

Why the CPU Family Matters

The CPU Family keeps the digital world alive:

- Babies make smart devices possible.
- Teens power our daily laptops.
- Parents keep professionals productive.
- Grandparents keep the internet running.
- Cousins give us pocket-sized supercomputers (phones).
- Family reunions tackle humanity's biggest problems.

From microwaves to supercomputers, the CPU Family proves that **every member matters**.

✨ Just like the **Bit Family** and **GPU Family**, this structure is:

- **Easy to understand** (family metaphors).
- **Layered by scale** (tiny → global).
- **Practical** (real-world examples).
- **Original writing** — I created these analogies fresh for your book.

Memory and Knowledge in AI

1. Memory

In humans: memory means recalling past experiences.
In AI: memory means the system can **remember past inputs, conversations, or events** and use them in the future.

- **Short-term memory:** Like a chatbot remembering what you said earlier in the same conversation.
- **Long-term memory:** Like AI storing knowledge from past training or user interactions and using it tomorrow, next week, or next year.

➡ Example: If your chatbot remembers that a student prefers video lessons over text, it can recommend videos automatically in the future.

2. Knowledge

In humans: knowledge means facts, skills, and understanding.
In AI: knowledge means everything the system has **learned from data** (during training, or through interaction).

- Stored in the network's **weights and connections** — this is how it recognizes patterns.
- Example: After being trained on thousands of math problems, the AI "knows" how to solve equations it has never seen before.

🌸 Plain English

- **Memory = remembering past experiences.**
- **Knowledge = what the AI has learned from all its data.**

For superintelligence, an AI would need *both*:

- The ability to **remember like a human** (memory).
- The ability to **understand the world deeply** (knowledge).

Introduction to ASCII

ASCII, short for **American Standard Code for Information Interchange**, is one of the earliest and most influential standards in computing. Before ASCII, different machines used different codes to represent text, making it difficult for computers to "talk" to each other. Introduced in the 1960s, ASCII solved this problem by assigning **numbers to characters**—letters, digits, punctuation marks, and control symbols—so that all computers could share and process text in a common way.

The ASCII system is built on a **7-bit code**, which allows for 128 unique values. For example, the capital letter **A** is represented by the number **65**, while the lowercase **a** is **97**. These values are stored as binary (0s and 1s), which computers can easily read and manipulate. Beyond letters and numbers, ASCII also includes **control characters** like "newline" or "tab" that tell computers how to format or manage text.

Although ASCII is limited to 128 characters, it became the **foundation of digital communication**. Later systems, such as **Unicode and UTF-8**, expanded on ASCII to support characters from languages all over the world, but they still keep ASCII as their base.

Understanding ASCII is like learning the alphabet of computers—it's the first step in how machines began to read, write, and exchange information

ASCII Values for Capital Letters (A–Z)

- A = 65
- B = 66

- C = 67
- D = 68
- E = 69
- F = 70
- G = 71
- H = 72
- I = 73
- J = 74
- K = 75
- L = 76
- M = 77
- N = 78
- O = 79
- P = 80
- Q = 81
- R = 82
- S = 83
- T = 84
- U = 85
- V = 86
- W = 87
- X = 88
- Y = 89
- Z = 90

ASCII Values for Lowercase Letters (a–z)

- a = 97
- b = 98
- c = 99
- d = 100
- e = 101
- f = 102
- g = 103
- h = 104
- i = 105
- j = 106
- k = 107
- l = 108
- m = 109
- n = 110
- o = 111
- p = 112
- q = 113

- r = 114
- s = 115
- t = 116
- u = 117
- v = 118
- w = 119
- x = 120
- y = 121
- z = 122

What Is Evolving AI?

Evolving AI is the concept of AI systems that **improve and adapt over time, almost like living things evolve.** Instead of staying fixed after being programmed, these systems can grow smarter by:

1. **Learning from Experience**
   - Just like humans learn from mistakes, evolving AI adjusts based on feedback.
2. **Adapting to New Situations**
   - Instead of only handling what it was originally trained for, it can deal with new problems.
3. **Self-Improvement**
   - Some evolving AI models can rewrite parts of their own "rules" or create new solutions, moving closer to true *general intelligence*.

Examples of Evolving AI

- **Chatbots:** Get better at conversations the more people interact with them.
- **Recommendation Systems (Netflix, YouTube):** Learn your habits and evolve to predict what you'll like next.
- **Robotics:** Robots that teach themselves to walk or pick up objects in smarter ways.

Why It Matters

- Evolving AI could lead us toward **Artificial General Intelligence (AGI)** — systems that think more like humans.

- It also raises questions about **ethics and control** — if AI keeps evolving, how do we guide it responsibly?

🌸 **Plain English:**
Evolving AI is simply AI that doesn't stay the same — it keeps learning, adapting, and improving, just like people (or nature) do.

📖 Understanding Tokens in AI (Word-to-Token Breakdown)

1. Average Word-to-Token Ratio

- In **English**, one token is about **¾ of a word** on average.
- Rule of thumb:
  - **100 tokens ≈ 75 words**
  - **1,000 tokens ≈ 750 words**
  - **4,000 tokens ≈ ~3,000 words** (about 6–7 pages of text).

2. Examples of Words and Tokens

- **Short words** (like "cat", "dog", "AI") → usually **1 token each**.
- **Longer words** (like "artificial" or "computerization") → may split into **2–3 tokens**.
  - Example: "computerization" → `computer` + `ization` = **2 tokens**.
- **Numbers**: "2025" → often **1 token**.
- **Punctuation**: Each symbol (like "?", "!") → **1 token**.
- **Spaces** → sometimes treated as tokens too!

3. Special Cases

- **Emojis** 🙂 → usually **1 token**.
- **Hashtags** (#AIStudyBuddy) → may split into 2–3 tokens depending on the model.
- **Foreign languages**:

- Chinese / Japanese → each **character = 1 token**.
- Korean → often split into syllables (so 1 word = 2–3 tokens).
-

4. Tokens in Practice (Real Examples)

Sentence:
👉 "AI helps people learn faster."

- Breakdown:
  - `AI` → 1 token
  - `helps` → 1 token
  - `people` → 1 token
  - `learn` → 1 token
  - `faster` → 1 token
  - `.` → 1 token

✅ Total = **6 tokens** (even though it's only 5 words).

5. Why Tokens Matter

- AI models (like GPT) don't think in words — they think in **tokens**.
- Each request you make to an AI is measured in tokens.
- The **context window** (memory limit) of the AI is in tokens:
  - GPT-3.5 → ~4,096 tokens (~3,000 words).
  - GPT-4 → up to 128,000 tokens (~100,000 words, or a short book!).

📝 Rule of Thumb for Your Readers

- **1 token ≈ ¾ word in English.**
- **1 page of text (double-spaced, ~300 words) ≈ 400 tokens.**
- **A book chapter (3,000 words) ≈ 4,000 tokens.**
-

◆ How Tokens Are Applied in AI

1. Tokenization (Breaking Text Into Tokens)

- The first step in applying tokens is **tokenization**: splitting text into smaller chunks the AI can handle.
- Example:
  Input: *"I love WordPress websites!"*
  Tokens: `[I] [love] [Word] [Press] [websites] [!]`

This allows the AI to process text at a granular level.

2. Mapping Tokens to Numbers

- Each token is converted into a **numerical ID** that represents it in the AI's "dictionary" (called a *vocabulary*).
- Example:
  - "Word" → 57321
  - "Press" → 44291
  - "websites" → 19583

The AI doesn't "see" words — it only works with numbers representing those tokens.

3. Feeding Tokens Into the Model

- Once tokenized and converted, the sequence of numbers is fed into the AI's **neural network**.
- Each token influences what comes next.
- The model predicts the most likely next token, one step at a time.

👉 Example in practice:
Input tokens = `[I] [love] [Word]`
AI predicts next token = `[Press]`.
Then next = `[websites]`.
Then `[!]`.

That's how AI generates coherent text.

4. Applying Tokens in Training

During **training**, millions or billions of tokens are fed into the model:

- Books, articles, websites, code → all turned into tokens.
- The model learns **patterns**: "What token usually follows what?"
- That's why it can later generate realistic answers.

5. Applying Tokens in Real Use

When you use AI (like ChatGPT, or a WordPress chatbot):

- **Input tokens** = your prompt/question.
- **Output tokens** = the AI's answer.
- The total tokens = cost + length limit.

👉 Example with OpenAI pricing:

- 1,000 tokens in input + 500 tokens in output = 1,500 tokens billed.

6. Application in Websites (Like WordPress)

When applying tokens inside an AI feature on your site:

- A visitor's question ("How do I reset my password?") is tokenized.
- Tokens are sent to the AI model (via an API).
- The AI processes them and returns output tokens ("Click 'Forgot Password'…").
- Those tokens are displayed as text on your site.

Everyday Analogy

Think of **tokens like Lego blocks**:

- Training = giving the AI billions of Lego blocks to learn shapes (language patterns).
- Using AI = giving it a handful of blocks (your prompt) and watching it build something new with them (the answer).

In short:
Applying tokens to AI means:

- **Breaking down text into tokens →**
- **Mapping tokens into numbers →**
- **Feeding them into the neural network →**
- **AI predicts the next tokens →**
- **Output tokens form the response you read.**

The Decimal Family

Another family we use every day is the **decimal family**, built from the digits 0 through 9. With a single digit, we can only count from 0 to 9. But when digits join together, the family grows: two digits (10–99), three digits (100–999), four digits (1,000–9,999), and so on. This family teaches us structure. It explains why people talk about "six-figure" or "seven-figure" incomes — they're simply counting how many digits are in the number. While the decimal family feels natural in English, in some languages it requires more effort, almost like using your fingers to carefully build the number. No matter the culture, the decimal family is how humans count — and it mirrors how computers count with bits.

Breakdown of the Decimal Family

- **1-digit numbers:**
  Range: `0 - 9`
  Example: Counting on your fingers.
- **2-digit numbers:**
  Range: `10 - 99`
  Example: Your age as a teenager, or the number of minutes in an hour (60).
- **3-digit numbers:**
  Range: `100 - 999`

Example: The number of pages in a small book, or the cost of a new smartphone ($699).

- **4-digit numbers:**
  Range: `1,000 - 9,999`
  Example: The year we live in (2025), or the population of a small town.
- **5-digit numbers:**
  Range: `10,000 - 99,999`
  Example: The price of a car ($25,000), or the number of people in a mid-size city.
- **6-digit numbers:**
  Range: `100,000 - 999,999`
  Example: A "six-figure income" ($100,000+), or the population of a large city.
- **7-digit numbers:**
  Range: `1,000,000 - 9,999,999`
  Example: A lottery jackpot ($5,000,000), or the number of followers a celebrity might have on social media.
- **8-digit numbers:**
  Range: `10,000,000 - 99,999,999`
  Example: The value of a company ($25,000,000), or the population of a large country's city.
- **9-digit numbers:**
  Range: `100,000,000 - 999,999,999`
  Example: A billion-dollar company (almost there), or the population of a big country.
-
- **10-digit numbers:**
  Range: `1,000,000,000 - 9,999,999,999`
-
- Example: One **billion**. Big tech companies like Apple and Google are worth over a trillion (12 digits).

💡 Why the Decimal Family Matters

The Decimal Family shows us **structure** in counting. Just like the Bit Family builds from 0s and 1s, the Decimal Family grows with digits. This is why people talk about **"six-figure" or "seven-figure" incomes** — it's simply about how many digits are in the number.

By understanding this ladder of digits, readers can connect **everyday life (money, years, population, social media followers)** with the same structure that computers use when measuring data in bits and bytes.

The **GPU**, or **Graphics Processing Unit**, is a type of computer chip originally designed to handle graphics and images. Its first purpose was to make video games look smooth by drawing millions of pixels on a screen very quickly. Over time, however, the GPU has grown into one of the most important processors in modern computing, far beyond just gaming.

Unlike a CPU, which is designed to do many different kinds of tasks one after another, the GPU is built for **parallel processing**. This means it can perform thousands of small calculations at the same time. For example, while a CPU might work like a skilled chef who prepares meals one by one, the GPU is like a team of hundreds of cooks all working together to make huge amounts of food at once. This structure makes the GPU perfect not only for graphics, but also for fields like artificial intelligence, data science, and scientific research.

Modern GPUs are used to train and run AI systems because they can process massive amounts of data in parallel. Training an AI model requires performing billions or even trillions of calculations, and GPUs can handle this workload much faster than CPUs alone. This is why companies such as NVIDIA, AMD, and Intel are competing to build more powerful GPUs for AI, cloud computing, and data centers.

In everyday life, GPUs still play a role in delivering smooth graphics for games, editing videos, or even powering the visuals on your smartphone. But in the AI world, GPUs are considered the **muscles of computation**—working alongside CPUs, which act more like the brains of the system. Together, they enable the powerful applications of modern artificial intelligence.

Types of GPU Memory

1. **GDDR (Graphics Double Data Rate)**
   - The most common type of memory in consumer and professional GPUs.
   - Versions like GDDR6 and GDDR6X offer high speeds for gaming, creative work, and moderate AI workloads.
2. **HBM (High Bandwidth Memory)**
   - Found in high-end AI and scientific GPUs.
   - Offers extremely high data transfer speeds by stacking memory chips vertically near the GPU core.
   - Versions include HBM2, HBM2e, and HBM3.
3. **Shared System Memory**
   - Used by integrated GPUs.
   - No dedicated VRAM; instead, the GPU uses a portion of the computer's main RAM.
   - Lower performance compared to dedicated memory.

Major GPU Manufacturers and Architectures

1. **NVIDIA**
   - Known for its CUDA architecture, which is widely adopted for AI, gaming, and professional graphics.
   - Product families include GeForce (gaming), RTX A-series (workstation), and H-series (data center and AI).
2. **AMD**
   - Uses the ROCm platform for AI and compute workloads, and RDNA architecture for gaming.
   - Produces Radeon (consumer), Radeon Pro (professional), and Instinct (data center) GPUs.

Intel

- A newer player in discrete GPUs, with the ARC series for consumers and specialized GPUs for data centers.
- Also a major supplier of integrated graphics in CPUs.

3. **Specialized AI Chip Makers**
   - Some companies focus solely on AI accelerators, which function similarly to GPUs but are purpose-built for machine learning.
   - Examples include Graphcore (IPU – Intelligence Processing Unit), Cerebras (Wafer Scale Engine), and Google (TPU – Tensor Processing Unit).

💡 **Educational Tip for Your Book Readers**:
While all GPUs are designed to process data in parallel, their performance depends on **core count**, **memory type**, and **intended workload**. Choosing the right GPU depends on whether you need it for **gaming, professional design, AI research, or mobile computing**.

The **CPU**, or **Central Processing Unit**, is often described as the *brain of the computer*. It is the part responsible for carrying out instructions from programs, whether that means doing calculations, moving data, or coordinating other components like memory and storage. Without the CPU, the rest of the computer would sit idle with no direction.

Inside the CPU, three main parts work together to make everything happen. The **control unit** acts like a traffic director, telling the system what to do next. The **arithmetic logic unit (ALU)** performs math and logical operations, while the **registers and cache** provide super-fast memory so the CPU can access important information quickly.

To understand this in everyday terms, think of a computer as a busy restaurant. The CPU is the **head chef** who takes in orders, organizes the kitchen, and makes sure dishes get served correctly. Other parts of the computer, like the GPU or system memory, are the kitchen staff and tools—but the chef is the one in charge.

The importance of the CPU is clear. A faster CPU allows programs to load more quickly and helps the system run smoothly, even when juggling many tasks at once. Modern CPUs often contain **multiple cores**, which means they can divide their workload and process different tasks simultaneously, making multitasking much easier.

In the world of AI, the CPU plays a supportive but essential role. While GPUs are usually favored for training large AI models—because they can handle thousands of calculations in parallel—the CPU is still responsible for organizing the process, managing data, and making sure everything runs in order. In short, the CPU is the foundation of computing, and even the most advanced AI would not function without it.

1. Types of CPUs by Purpose

1. **Desktop CPUs**
   - Designed for personal computers used in homes and offices.

- Balance speed, power efficiency, and cost.
- **Examples**: Intel Core series, AMD Ryzen series.

2. **Laptop CPUs (Mobile Processors)**
   - Built for lower power consumption to extend battery life.
   - Usually run at lower clock speeds than desktop versions but are more energy efficient.
   - **Examples**: Intel Core mobile processors, AMD Ryzen Mobile series, Apple M-series.
3. **Server CPUs**
   - Designed for high-performance, 24/7 workloads in data centers.
   - Support many cores, large memory capacity, and advanced virtualization.
   - **Examples**: Intel Xeon, AMD EPYC.
4. **Embedded CPUs**
   - Used in devices like routers, industrial machines, and IoT devices.
   - Built for reliability and low power usage rather than raw speed.
   - **Examples**: ARM Cortex-M series, Intel Atom.

5.Smartphone/Tablet CPUs (SoCs)

- System-on-a-Chip designs that integrate the CPU with GPU, memory controllers, and other components.
- **Examples**: Qualcomm Snapdragon, Apple A-series, Samsung Exynos.

6 Types of CPUs by Architecture

1. **x86 / x86-64 Architecture**
   - Found in most desktops, laptops, and servers.
   - Supports a wide range of operating systems and applications.
   - **Examples**: Intel and AMD processors.
2. **ARM Architecture**
   - Popular in mobile devices and increasingly in servers for energy efficiency.
   - Known for low power consumption.
   - **Examples**: Apple M1/M2/M3, Qualcomm Snapdragon, AWS Graviton.
3. **RISC-V Architecture**
   - Open-source CPU architecture that's gaining popularity for custom designs.
   - Used in research, IoT devices, and emerging computing platforms.

7.Types of CPUs by Core Design

1. **Single-Core CPUs**
   - One processing core — rare today except in basic embedded devices.
   - Processes one task at a time.
2. **Multi-Core CPUs**
   - Two or more cores for handling multiple tasks in parallel.

    - Common configurations: dual-core, quad-core, hexa-core, octa-core, and beyond.
3. **Hybrid-Core CPUs (Big.LITTLE design)**
    - Mix of high-performance cores and power-efficient cores in the same chip.
    - Balances performance with energy savings.
    - **Examples**: Intel Alder Lake architecture, ARM big.LITTLE CPUs.

8. Types of CPUs by Form Factor

1. **Desktop Socketed CPUs**
    - Removable chips that fit into a CPU socket on a motherboard.
    - Easier to upgrade.
2. **Mobile Soldered CPUs**
    - Permanently attached to the motherboard in laptops and small devices.
    - Not user-upgradable.
3. **SoCs (System on a Chip)**
    - CPU integrated with GPU, RAM controller, and other functions.
    - Common in mobile devices and small embedded systems.

💡 **Educational Insight for Your Readers**:
While all CPUs process instructions, their design depends heavily on the **type of workload** they're built for. For example:

- A **server CPU** might have 96 cores for cloud computing.
- A **mobile CPU** might have fewer cores but focus on battery life.
- An **embedded CPU** may run for years with minimal power in industrial machines.

CPU vs GPU – Key Differences

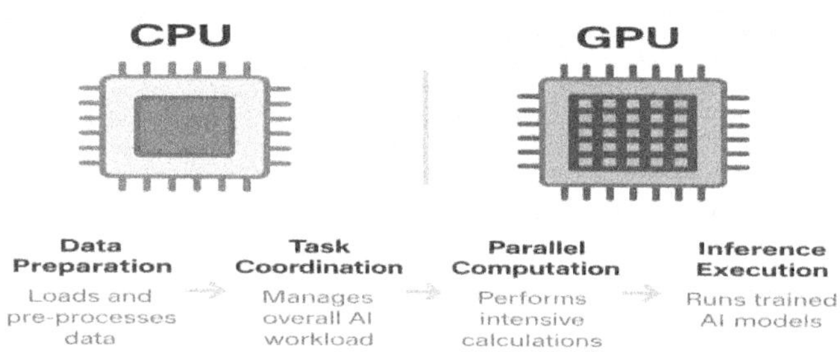

Feature	CPU (Central Processing Unit)	GPU (Graphics Processing Unit)

Feature	CPU (Central Processing Unit)	GPU (Graphics Processing Unit)
Main Role	General-purpose computing; executes a wide variety of tasks and instructions.	Specialized for parallel processing; handles large amounts of data simultaneously.
Core Count	Fewer, more powerful cores (e.g., 4–32 cores in most systems).	Hundreds or thousands of smaller, efficient cores.
Processing Style	Optimized for sequential tasks — fast at single-thread performance.	Optimized for parallel tasks — processes many operations at once.
Workload Types	Operating systems, office software, web browsing, most standard applications.	Graphics rendering, video processing, AI training, simulations, large-scale data analytics.
Memory Type	Uses system RAM (DRAM) connected via memory controllers.	Uses dedicated video memory (VRAM) with high bandwidth, often GDDR or HBM.
Clock Speed	Generally higher (3–5 GHz typical), prioritizing quick execution per core.	Usually lower per core, but massively parallel, prioritizing throughput.
Power Consumption	Moderate to high, depending on workload.	High, especially for data center or AI GPUs.
Flexibility	Very versatile; can handle almost any computing task.	Highly specialized; best at tasks that can be broken into parallel workloads.
AI Role	Coordinates tasks, handles logic, runs parts of AI inference that aren't GPU-optimized.	Trains and runs AI models quickly by processing vast amounts of matrix math in parallel.

Educational Note:

- The **CPU** is like the *brain's decision-maker* — good at reasoning and complex logic.
- The **GPU** is like a *massive team of workers* — not as smart individually, but extremely fast when working together on repetitive jobs.
- In AI systems, **CPUs and GPUs work together**: the CPU prepares and organizes data, while the GPU does the heavy number-crunching.

TensorFlow – Quick Summary

TensorFlow is a **free and open-source platform** designed to help developers and researchers create, train, and deploy machine learning and deep learning models. Originally developed by the **Google Brain** team, it offers a full set of tools and libraries for building AI systems that can run on devices ranging from laptops and servers to mobile phones and cloud infrastructure.

In-Depth Look at TensorFlow

1. **Who Created It**
   - Developed by **Google Brain** and first released to the public in 2015.
2. **Main Goal**
   - To simplify the process of creating AI models that can operate across different environments — from small personal projects to large-scale production systems.
3. **Languages It Works With**
   - Primary support is for **Python**, but APIs are also available for **C++**, **JavaScript**, **Java**, **Go**, and other languages.
4. **Notable Features**
   - **Data Flow Graphs**: Organizes computations into a graph structure for efficiency and scalability.
   - **Multi-Platform Support**: Can run on CPUs, GPUs, and specialized processors like Google's TPUs.
   - **Scales Up or Down**: Works on a single device or across multiple servers in distributed setups.
   - **GPU Acceleration**: Integrates with NVIDIA's CUDA technology for faster processing.
   - **Keras Integration**: Offers an easy-to-use, high-level API for quickly building and testing models.
5. **Main Components**
   - **TensorFlow Core**: The foundational library for building and running models.
   - **TensorFlow Lite**: Optimized for mobile devices and embedded systems.
   - **TensorFlow.js**: Lets you run AI models directly in a web browser.
   - **TensorFlow Extended (TFX)**: A suite for deploying and managing AI workflows in production.
   - **TensorFlow Hub**: A shared repository of pre-trained models ready for use.
6. **Where It's Used**
   - Applications include computer vision (image and video recognition), speech-to-text systems, natural language processing, predictive analytics, recommendation engines, and generative AI.
7. **Benefits**
   - Large and active global community.
   - Detailed documentation and tutorials.
   - Access to many pre-built models for quick experimentation.
   - Highly flexible for both beginners and advanced AI developers.
8. **Challenges**
   - Can be more complex to learn than simpler ML libraries.
   - Performance tuning may require deeper technical knowledge.
9. **License Information**
   - Distributed under the **Apache License 2.0**, which means it's free to use, adapt, and share.

💡 **Educational Insight**:
Think of TensorFlow as a **construction kit for AI** — it gives you the framework, building

blocks, and tools needed to create anything from a simple chatbot to an advanced neural network powering self-driving technology.

AI Chip Power Consumption refers to the amount of electrical energy that computer chips, such as CPUs, GPUs, and specialized AI processors, use while performing tasks. Power consumption is usually measured in **watts (W)**, which tells us how much energy the chip draws at any given moment. The more powerful the chip and the heavier the workload—such as training large AI models—the more electricity it requires. For example, a modern GPU used for AI training can consume **hundreds of watts**, and when thousands of these chips run together in a data center, the total energy demand can reach into the **megawatts**, similar to powering an entire town. High power consumption not only increases electricity costs but also generates heat, which means additional cooling systems are needed. In the world of AI, understanding chip power consumption is important because it directly affects efficiency, cost, and environmental impact.

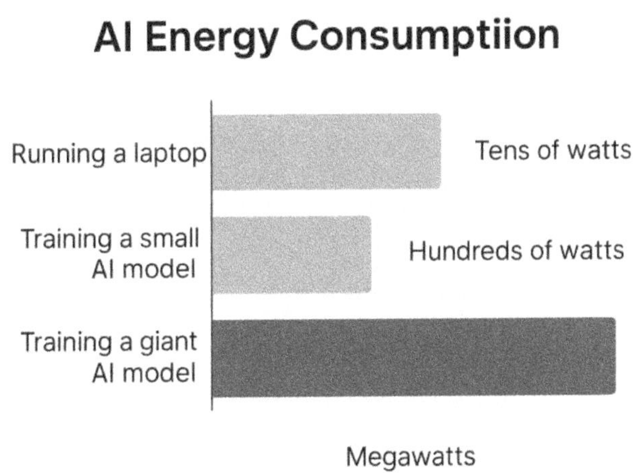

1. AI Chip Power Consumption

- **Individual GPU/Accelerator**
  - High-end AI chips like the NVIDIA H100 typically use **300–700 watts** each.
  - Clusters for training large models can include **thousands** of GPUs running continuously.
- **Example**:
  - 1 H100 at 400 watts running 24 hours = **9.6 kWh** per day.
  - 1,000 H100 GPUs = **9,600 kWh per day** — enough to power hundreds of homes for a day.

2. Data Center Overhead

- **Cooling Systems**
  - Chips produce massive heat; advanced cooling (air or liquid) can use **30–50%** as much power as the computing hardware itself.
- **Networking Equipment**
  - High-speed interconnects like InfiniBand switches and routers draw additional power.
-
- **Storage Systems**

SSD arrays, backup systems, and data pipelines also run 24/7.

Example Ratio:
If GPUs use 10 MW of power, cooling and other infrastructure may add **3–5 MW more**.

3. Operational & Support Systems

- **Lighting, security systems, monitoring equipment**
- **Backup power systems** like diesel generators or battery storage
- **Power conversion losses** (some energy is lost converting AC power from the grid into the DC power chips use)

Putting It Together – A Large AI Training Run

Let's take a hypothetical **training run for a GPT-scale model**:

1. **Compute Hardware**:
   - 2,000 GPUs × 500 W each = 1 MW.
2. **Cooling Overhead**:
   - Adds another ~0.5 MW.
3. **Other Infrastructure**:
   - Networking, storage, and support systems = ~0.2 MW.

Total Power Draw = 1.7 MW (continuous).
If this training runs for **14 days straight**:

- 1.7 MW × 24 hours × 14 days = **571 MWh** (megawatt-hours) — enough to power roughly **50 U.S. homes for a year**.

💡 **Educational Note**:
The energy footprint of AI is growing fast — not just because of bigger models, but also because each new generation of chips is more powerful, allowing companies to train **larger models more often**.

The AI chip landscape is not dominated by a single company but by a **network of innovators and partners** working at different layers of the industry. Designers create the architecture, foundries bring those blueprints to life, memory suppliers enable massive data handling, and startups test bold new ideas. This collaboration fuels the rapid growth of artificial intelligence, from powering everyday smartphones to training the largest AI models in the world. For readers, understanding these players reveals why the "chip rush" is often compared to a digital gold rush — because whoever controls the chips ultimately helps shape the future of technology.

1. Leading AI & GPU Chip Companies

These companies design chips specifically for **AI training, GPUs, and data center acceleration**:

- **NVIDIA** – Leader in AI GPUs (A100, H100) and CUDA platform.
- **AMD** – Produces Instinct MI series for AI and Radeon GPUs.
- **Intel** – CPUs, integrated graphics, and AI accelerators (Gaudi, Habana Labs).
- **Google** – Builds custom **TPUs** (Tensor Processing Units) for in-house and Google Cloud AI.
- **AWS (Amazon Web Services)** – Custom AI chips like **Inferentia** and **Trainium** for cloud workloads.
- **Apple** – M-series processors with built-in Neural Engines for AI on devices.
- **Meta** – Designing custom AI chips for internal use (MTIA – Meta Training and Inference Accelerator).
- **Tesla** – Dojo AI training chip for self-driving technology.

2. Foundries & Manufacturing Giants

These companies **fabricate** chips for many AI designers:

- **TSMC (Taiwan Semiconductor Manufacturing Company)** – Manufactures chips for NVIDIA, AMD, Apple, and more.
- **Samsung Electronics** – Chip manufacturing and memory production.
- **Intel Foundry Services** – Intel's own manufacturing division.
- **GlobalFoundries** – Supplies chips for automotive, IoT, and industrial AI.

3. Memory & Storage Chip Leaders

AI hardware depends heavily on **high-bandwidth memory**:

- **Micron Technology** – Makes GDDR and HBM memory for GPUs.
- **SK Hynix** – Supplies HBM for NVIDIA's H100 GPUs.
- **Samsung** – Major supplier of both HBM and GDDR memory.

4. Specialized AI Chip Startups

Focused entirely on **custom accelerators** for AI:

- **Cerebras Systems** – Known for the **Wafer-Scale Engine** (largest chip ever built).
- **Graphcore** – Intelligence Processing Unit (IPU) for AI workloads.
- **SambaNova Systems** – Dataflow architecture for AI model training.
- **Tenstorrent** – Led by ex-AMD engineers, building AI and RISC-V chips.

💡 **Educational Note**:
The "chip rush" isn't just about **one company** — it's an ecosystem.

- **Design companies** (like NVIDIA, AMD, Apple) create the blueprints.
- **Foundries** (like TSMC, Samsung) physically manufacture the chips.
- **Memory suppliers** (like Micron, SK Hynix) build the high-speed memory needed for AI.
- **Startups** push the limits with experimental chip designs.

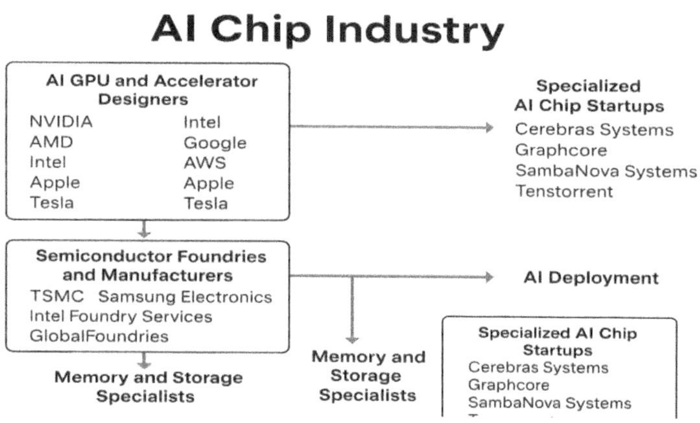

Table of Contents

Part I – Getting Started

Part II – Building the Store

Part III – Adding AI to the Site

Part IV – Growing and Polishing

1.Introduction:

Welcome to *No Code, No Problem: Build Your Website & Chatbot with WordPress & Botpress*—a book born out of determination, curiosity, and the dream of building something from the ground up without any formal tech background.

If you've ever felt overwhelmed by terms like *web hosting*, *plugins*, *API*, or *chatbot integration*, you're in the right place. This book isn't written by a software engineer or web developer it's written by someone just like you. Someone who had a vision, rolled up her sleeves, and figured it out one step at a time.

When I started building my own website, **YourAIStudyBuddy.com**, I wasn't studying computer science, nor did I have a team of tech experts behind me. I was running a flower shop, armed with a passion for learning, a stack of notebooks full of plans, and a dream to create something valuable—an educational and e-commerce platform that could help others learn about Artificial Intelligence in a simple, approachable way.

In the past, I had hired freelancers from Fiverr to build websites for me. But this time, I didn't want to hand the project over and walk away. I wanted to understand it, shape it, and eventually teach others how to do it too.

In this book, I take you behind the scenes of building a full-fledged WordPress e-commerce website—from registering a domain name to selecting a hosting provider (in my case, Hostinger), connecting WordPress.org, setting up Stripe for payment processing, adding digital products like books and videos, and integrating an AI chatbot built on Botpress. I'll share what worked, what didn't, the roadblocks I faced, and how I overcame them—not with coding, but with creativity, focus, patience, and a lot of Google searches.

What makes this guide different is that it's not theory--it's practical, personal, and packed with real-world experience. Each chapter is a building block that takes you from zero to launch—no coding skills required. Whether you're an artist, educator, business owner, or someone with a big idea and no tech background, this book will give you the roadmap you need to get started.

You'll learn how to:

1. Choose and register your domain name
2. Set up WordPress and pick the right theme and plugins
3. Structure your pages and connect all the moving parts
4. Add secure payment systems like Stripe
5. Install and personalize your own AI chatbot
6. Get your website indexed on Google so it appears in search results
7. Launch with confidence and clarity

This is more than just a book about websites it's a book about believing in yourself. If I can do it, so can you. You don't need to be tech-savvy--you just need to be committed..Let's begin this journey together.

Chapter 1: I Didn't Know What I Was Doing

I have never built a website in my life. Let me repeat that: I have never—not once—created a website from scratch.

Sure, I worked around websites when I was selling them through AT&T. But let's be honest—selling websites and actually building one are two very different things. I could talk about them, close deals, and meet quotas. But when it came to logging in, designing, or connecting anything technical? I was completely lost.

When I first heard about WordPress.org, I had no idea what it was. In fact, I didn't even know there were two WordPress platforms—WordPress.org and WordPress.com. It was confusing and honestly frustrating. One is self-hosted, and the other is hosted for you. But when you're brand-new and non-technical, they might as well be written in another language.

I wish someone had just explained it simply:

1. **WordPress.org** gives you full control over your site—but you're also responsible for managing it.
2. **WordPress.com** is like training wheels—more limited, but easier to start with.

I knew what a domain name was—it's the website address people type in. But I didn't know where to check if a name was available, how to buy one, or what to do once I had it. When I came up with **YourAIStudyBuddy.com**, I had no idea how to "claim" it or point it anywhere. It felt like trying to buy land in a country where I didn't speak the language. Terms like **domain registrar**, **DNS settings**, and **name servers** sounded like tech talk from another planet.

Then came WordPress itself. When I logged into my dashboard for the first time, I just stared at the screen. Where was I supposed to go? What did all these tabs mean? And don't get me started on the hidden menus. For example, "Tools" isn't under "Settings" where you might expect—it's buried elsewhere. I clicked around nervously, hoping I wouldn't break anything. It felt like wandering through a maze where every turn led to another hallway of confusion.

Just when I thought I was making progress, I met another new word: **hosting**. **Hostinger**, my chosen provider, entered the picture—and I quickly learned that WordPress.org needs a hosting platform to actually run your website. Connecting my domain name to Hostinger, installing WordPress, and setting up the back end felt like a giant puzzle where I didn't have all the pieces. Their dashboard wasn't always intuitive, and I spent hours watching tutorials, reading articles, and still feeling stuck.

There were moments I wanted to cry—honestly cry. I sat in front of my screen with blurry eyes, my head pounding, and a deep sense of frustration. I wasn't lazy. I wasn't unmotivated. I was just overwhelmed. Too many hidden menus, too many logins, too many steps that didn't make sense unless you were already "in" the world of websites.

3.But somewhere in that chaos, something shifted. I stopped trying to "know everything" and started focusing on learning one small thing at a time.

I learned how to log in properly.
I learned what a plugin is.
I learned the difference between a theme and a widget.

I didn't need to be a tech wizard—I just needed persistence. And that persistence became my greatest strength.

Each time I figured out one more piece—whether it was installing a plugin, updating a theme, or setting up a contact form—I felt more confident. And then, one day, I looked at my screen and realized something: I was doing it.

I had gone from knowing nothing to creating something. My AI learning website wasn't just an idea anymore—it was real.

Key Takeaway
You don't need to know everything at once. Start with small steps:

1. Learn one new term or function at a time.
2. Don't be afraid to click around—mistakes are part of learning.
3. Persistence will take you further than technical skills at the start.

If I can go from zero experience to launching a full AI learning platform, so can you.

Chapter 2: Spending Time on Learning the WordPress Dashboard & Building from the Back End

After my emotional meltdown — and trust me, it was a real one — I knew I had to slow down and truly understand what I had gotten myself into. I had officially stepped into the world of **WordPress.org**, and every click felt like walking into a maze with hidden doors.

I didn't know where anything was, and part of me wanted to give up right then. But another part of me whispered, *"You've come this far. Take it one step at a time."*

So that's what I did.

I decided to dedicate real, focused time to learning the WordPress dashboard. Not a quick five minutes here and there, but hours where I could truly explore without rushing. That decision became the turning point.

At first, I didn't even know it was called the *dashboard.* I just saw a dark menu running down the left-hand side of my screen. What I didn't realize was that every label, every section in that 4.menu — *Posts, Pages, Media, Appearance, Plugins, Settings* — was part of my **control center**. This wasn't just decoration. This was the brain of my website.

The first "small mountain" I had to climb was simply **logging in**. Sounds basic, right? But even that wasn't smooth. I learned quickly that you need:

1. A valid email (if it's wrong or inactive, you can't even start).
2. A strong password that's saved somewhere safe (otherwise, you'll lock yourself out before you begin).

I made both mistakes more than once.

When I finally got in, the dashboard felt overwhelming. It was like stepping into a giant warehouse with no signs telling me where to go. But slowly, I started to understand that each section had its own purpose. I wasn't just randomly clicking anymore — I was learning what each tool did.

And that's when it clicked: **a WordPress website is built from the back end first**.

The beautiful homepage you eventually see? That comes later. The back end is where the magic — and the heavy lifting — happens. It's where you make decisions about your site's structure, where you install plugins, create pages, and lay down the foundation before adding the "pretty" things like colors, logos, and images.

It's the difference between decorating a house and actually building the house.

Before you can show off your site to the world, you need that strong foundation. That means learning your way around the back end of WordPress.org — and that's exactly what I did next.

In the following chapters, I'll walk you through exactly how I:

1. Set up my site's back end
2. Created my first pages
3. Installed the right plugins
4. Built a structure strong enough to hold all the beautiful content I wanted to share

I'll explain it in plain language so you don't feel lost the way I did. Because if I can go from total beginner to confident website builder, you can too.

Looking back, I'm proud of how far I've come. I'll never forget those early weeks of confusion — but they taught me persistence. That's why I'm writing this book: to help people who are smart, determined, and capable, but simply need someone to guide them through the tech jungle.

Key Takeaway

Before you focus on making your website "look good," make sure you understand how it's **built**. The back end is your foundation — and without it, everything else will crumble. Take the time to explore the dashboard, learn where everything is, and practice until it feels natural.

WORDPRESS DASHBOARD OVERVIEW

The WordPress Dashboard is the control center of your entire website.

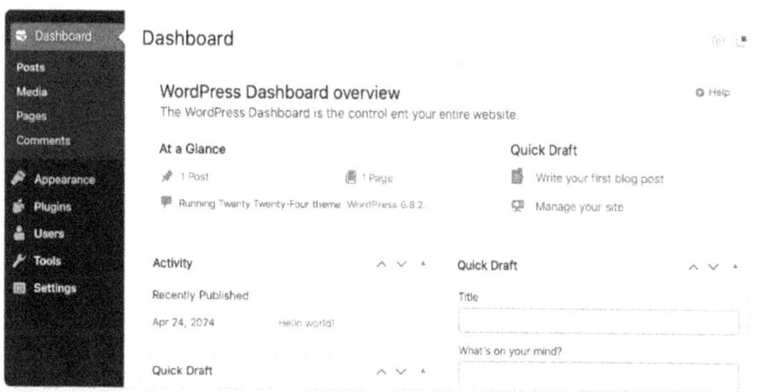

The image below shows a real-life snapshot of the Dashboard so you can get familiar with the look and feel of your workspace before you begin building.

Chapter 3: Why It's Important to Know the Backend of Your Website

When most people picture "building a website," they think of colors, images, menus, and the pages visitors see. That's called the **frontend** — the public face of your site.

But behind every beautiful website is the **backend** — the control room where everything is created, organized, and maintained. If the frontend is the stage, the backend is the director's booth, where the lights, sound, and scene changes are controlled.

At first, the backend can feel intimidating. I remember staring at the WordPress dashboard, wondering what all the menus meant. But once I understood it, everything changed. Knowing how to use the backend — even just the basics — gave me power, independence, and confidence.

The Backend Is Your Control Center

This is where you:

1. Create pages
2. Write blog posts
3. Upload images
4. Approve or delete comments
5. Install plugins and themes

6. Connect email tools
7. Add payment systems
8. Set up your menus
9. Secure your site

If you don't know how to navigate the backend, it's like trying to drive a car without knowing where the steering wheel or brakes are.

Why It Matters

1. It Gives You Freedom and Control:

When you understand the backend, you're no longer at someone else's mercy for small updates. You can:

- Update your homepage instantly
- Fix broken links yourself
- Add new products
- Create contact forms
- Manage subscribers

This means you can keep your site fresh without waiting — or paying — for someone else to do it.

2. It Saves You Money:

Hiring a web developer for every small change adds up quickly. By learning the backend, you can handle 90% of your site's day-to-day needs yourself and only hire help for big technical jobs.

3. It Builds Your Confidence : At first, menus and settings seem like they're written in another language. But the more time you spend exploring, the more familiar it becomes. Eventually, that fear is replaced with: *"Wow... I can actually do this."*

Main Sections of the WordPress Backend

When I first opened the dashboard, I had no idea what most of these did. To save you from that same confusion, here's a simple guide:

- **Dashboard**
  - *What it does:* Your homepage for site management. Shows quick updates, comments, plugin notices, and site health.
  - *Why it's important:* Lets you see what needs attention immediately.
- **Posts**
  - *What it does:* Create and manage blog posts with categories, tags, and images.
  - *Why it's important:* Builds your content library and helps with SEO.
- **Media**
  - *What it does:* Stores all your images, videos, PDFs, and audio.
  - *Why it's important:* Keeps all your visuals in one organized place.
- **Pages**
  - *What it does:* Creates key site pages like Home, About, Contact, or Products.

    - *Why it's important:* These are the backbone of your website's structure.
- **Comments**
    - *What it does:* Shows feedback on posts or products for you to approve, delete, or reply to.
    - *Why it's important:* Helps you engage with your audience and moderate discussions.
- **Appearance**
    - *What it does:* Manages your site's design — themes, menus, widgets, homepage layout.
    - *Why it's important:* This is where you define your brand's look.
- **Plugins**
    - *What it does:* Adds extra features to your site, like eCommerce, SEO tools, and chatbots.
    - *Why it's important:* Turns a basic site into a feature-rich platform.
- **Users**
    - *What it does:* Manages who can log in and what they can do (Admin, Editor, Subscriber, etc.).
    - *Why it's important:* Crucial for team or multi-author sites.
- **Tools**
    - *What it does:* Offers import/export features and plugin-specific tools.
    - *Why it's important:* Useful for moving data or doing advanced setup tasks.
- **Settings**
    - *8 What it does:* Manages your site's title, URL, time zone, language, permalink structure, and more.
    - *Why it's important:* One of the most important areas for making sure your site runs correctly.

Extra Menu Items After Installing Plugins

When you install certain plugins, new menu options will appear. For example:

- **WooCommerce** – Manage products, orders, and payments.
- **Products** – List and edit store items.
- **WP Mail SMTP** – Email configuration.
- **Botpress Chat** – Chatbot settings.
- **Stripe** – Payment gateway setup.

It Prepares You for Growth

Whether you're blogging, selling products, teaching courses, or sharing videos, you'll need to know your backend to:

- Connect email marketing tools
- Add subscription plans
- View user data
- Back up your site

The bigger your site grows, the more you'll rely on backend skills.

Key Takeaway

Your backend is the **engine of your website**. Learn it, and you gain control, independence, and confidence. Skip it, and you'll always depend on someone else to "drive" your site.

Chapter 4: Creating Your Website Pages

(Home Page, Login Page, Products Page, Terms & Conditions, and More)

When you think of building a website, one of the first things to understand is **Pages**. Pages are like the permanent **rooms in your website** — each one has a specific purpose.

- **9. Home Page** – The welcome area where visitors first arrive.
- **Login Page** – A secure entryway for members or customers.
- **Products Page** – The showroom where you display your items or services.
- **Terms & Conditions** – The rulebook for how visitors can use your site.

Unlike blog posts, which are time-stamped and can get buried over time, pages are **static and long-lasting**. They're the foundation of your site's structure.

💡 **Pro Tip:** Imagine walking into a store with no clear signs or labels. Confusing, right? Your pages are those signs — they guide visitors to exactly what they need.

Why Pages Matter

Without well-organized pages:

1. Visitors might get lost and leave quickly.
2. Search engines may not list you correctly.
3. You could face legal or trust issues if key pages (like Terms & Conditions) are missing.

Your pages are the **storefront windows** of your site — they must be clean, clear, and inviting.

Pages vs. Posts — What's the Difference?

This is a common confusion for beginners.

- **Pages** = Permanent, static content (Home, About Us, Products, Terms & Conditions).
- **Posts** = Time-stamped, frequently updated content (blog articles, announcements, news).

Think of pages as the walls and rooms of your house, and posts as the notes you pin on the fridge. Both are important, but they serve different purposes.

Where to Find Pages in WordPress

1. Log in to your WordPress dashboard.

2. On the left menu, click **Pages → Add New**.

Step-by-Step: Creating a Page

Step	Action	Beginner Tip
1	Click **Add New Page**	Opens the Block Editor — your blank canvas.
2	Give your page a short, clear title.	Example: "Home" or "Products."
3	Add text, images, or buttons.	Use headings (H2, H3) for readability.
4	Set a Featured Image (optional).	May appear in previews or social shares.
5	Assign a Parent Page if needed.	Example: "FAQ" under "Support."
6	Choose visibility: Public, Private, or Password Protected.	Useful for member-only areas.
7	Add SEO info (title tag & meta description).	Use plugins like Yoast SEO or Rank Math.
8	Save, Preview, Publish, and View.	Always check before going live.

Save, Preview, Publish, and View — Your Safety Net

These four buttons are your best friends in WordPress:

1. **Save Draft** – Keeps your work private until you're ready.
2. **Preview** – Shows how the page will look before it's public.
3. **Publish** – Makes your page live. You can also schedule a future date.
4. **View** – Double-check the live version after publishing.

💡 **Always check both desktop and mobile views** — most visitors are on their phones.

Hidden Layers to Master

Once you're comfortable with basic pages, here are some extra details worth knowing:

- **Layout choice:** Block Editor vs. Classic Editor.
- **Menu placement:** Decide if the page will appear in navigation.
- **SEO setup:** Write clear meta descriptions.
- **Accessibility:** Add alt text to all images.

Avoid These Mistakes 🚫

1. Forgetting to check mobile view.
2. Publishing placeholder text or images.
3. Not adding the page to your navigation menu.
4. Skipping SEO descriptions.

Quick Checklist ✅

Before you hit publish, make sure:

1. Title is short and clear.
2. Content is formatted with headings.
3. Images are optimized with alt text.
4. SEO title & meta description are set.
5. Page is added to your menu.
6. Tested on both desktop and mobile.

Key Takeaway

Pages are the **permanent structure** of your website. Build them with care, keep them organized, and always test before publishing. If posts are your conversation with visitors, pages are your foundation. Without strong pages, the rest of your website can't stand.

Chapter 5: Writing Blog Posts That Attract and Engage

When I first heard the word *blog post*, I didn't fully understand what it meant. Was it the same as a page? Was it just an article? The truth is, blog posts are very different from your website pages — and once I learned that difference, everything started to make sense.

Blog posts are articles you add to your site to share updates, tutorials, stories, or opinions. They're time-stamped and typically appear in reverse chronological order — meaning your newest post is shown first.

Some examples include:

- Latest store or business news
- "How-to" tutorials for your products or services
- Special announcements or promotions
- Industry insights or commentary
- Behind-the-scenes updates

Unlike static pages (like your Home or About page), blog posts are **dynamic**. They keep your site fresh and give your visitors a reason to come back.

Why Blog Posts Matter

1. **Boost Your SEO** – Google loves websites that update regularly. Each new post is another chance to rank for keywords in your niche.
2. **Build Authority** – Sharing valuable content positions you as an expert in your field.
3. **Engage Your Audience** – Posts give visitors a reason to interact with you.
4. **Drive Sales** – Well-placed calls-to-action in your posts can guide readers to products, services, or sign-ups.

💡 **Pro Tip:** Even one helpful post per month can gradually improve your Google ranking and build audience trust.

My First Blog Post (and What Went Wrong)

When I wrote my very first blog post, I thought I had done everything right. I poured my heart into it, clicked publish, and checked it on my phone. What I saw shocked me — one giant wall of text with no headings, no images, and no way for the reader to breathe.

That was my lightbulb moment: **blog posts aren't just about words — they're about presentation.** From that point forward, I started using short paragraphs, bullet points, and images to break things up. It instantly made my writing easier to read, and my readers stayed on the page longer.

Where to Find Posts in WordPress

1. Log in to your WordPress Dashboard.
2. In the left-hand menu, click **Posts → Add New.**

Step-by-Step: Writing a Blog Post

Step	Action	Beginner Tip
1	Click **Add New Post**	Choose a title that's descriptive and engaging. Example: "5 Easy Ways to Protect Your Website."
2	Write Your Content	Start with an intro, break into short paragraphs, and use H2/H3 headings.
3	Add Images or Videos	Compress files (use TinyPNG or Squoosh) so your site loads faster.
4	Assign Categories & Tags	Categories = broad topics, Tags = specific details.
5	Write an Excerpt (Optional)	A short summary for previews.
6	Choose Publish Date or Schedule	Scheduling helps with consistency.
7	Preview, Then Publish	Always check on both desktop and mobile before going live.

Example Blog Post Structure

Title: *7 Ways to Keep Your WordPress Site Secure*
Intro: Start with a hook — why security matters.
Body: Use 3–7 main points, each with its own heading and explanation.
Call-to-Action: Invite readers to comment, share, or click a product/service link.
Conclusion: Recap main points and leave them with a final thought or question.

Formatting for Readability

- Use **short paragraphs** (2–4 sentences).
- Break text with **headings**.

- Add **bullet points or numbered lists**.
- Highlight important terms with **bold or italics**.
- Always check **mobile view** — most people will read on their phone.

Optimizing Blog Posts for SEO

1. Install an SEO plugin (Yoast SEO or Rank Math).
2. Target 1–2 primary keywords per post.
3. Add your keyword to:
   - Title
   - First paragraph
   - At least one subheading
   - Meta description
4. Add **internal links** to other posts or pages.
5. Use **alt text** for all images.

💡 **Pro Tip:** Write for humans first, search engines second. Don't "stuff" keywords.

Hidden Features Worth Knowing

- Embed videos from YouTube or Vimeo by pasting the link.
- Link to related posts to keep visitors on your site longer.
- Use a **Featured Image** to make your post stand out in previews and social shares.

Avoid These Mistakes 🚫

1. Publishing without proofreading.
2. Forgetting categories and tags.
3. Uploading huge, uncompressed images.
4. Writing long walls of text with no headings.

Quick Checklist ✅

- Catchy, clear title.
- Content is well-formatted with headings.
- Images optimized and alt text added.
- Categories and tags assigned.
- SEO title & meta description set.
- Internal links included.
- Post scheduled or published.
- Previewed on mobile and desktop.

Final Takeaway

Blogging is one of the most effective ways to grow your audience, improve SEO, and build trust. The key isn't perfection — it's **consistency**. Even one quality post per month can have a big impact over time.

If I can go from publishing a messy wall of text to writing posts that actually attract readers, you can too. Start simple, keep learning, and keep writing.

Chapter 6: Your Website's Control Room — Backend Essentials for Growth

When I first stepped into the backend of my WordPress site, I felt like I had opened the door to a spaceship cockpit. Buttons, menus, and settings were everywhere, and I had no idea what any of them did. My biggest fear? Clicking the wrong thing and "breaking" my site. But over time, I realized the backend wasn't something to avoid — it was the beating heart of my website. Once I learned how to navigate it, I gained control, freedom, and confidence.

The backend is your website's **control room**. While the frontend is what the world sees — the design, the colors, the menus — the backend is where all the real decisions are made. It's where you create, organize, and secure everything that keeps your site running.

Why Backend Knowledge Matters

1. It Puts You in Control

From one dashboard, you can:

- Update pages and publish new posts
- Add products or services
- Change design elements
- Manage payments and orders
- Monitor site activity

At first, it looks intimidating — like sitting in a pilot's cockpit with hundreds of switches. But once you start exploring, you realize most of those "buttons" are just tools waiting to be used. The more you practice, the more natural it feels, and soon enough, you'll be flying the plane yourself.

2. It Saves You Money

I'll never forget the first time I updated my homepage by myself. It took hours of trial and error, but when I finally figured it out, I realized something important: I didn't need to pay someone for every little change. When you understand the backend, you only hire developers for big projects — not everyday tasks. That alone can save you hundreds or even thousands of dollars over time.

3. It Builds Confidence

In the beginning, every click felt risky. But with practice, that fear turned into confidence. Now, instead of thinking, *"What if I break something?"* I think, *"I know exactly where to go."* That shift in mindset was one of the most empowering parts of my journey.

4. It Prepares You for Growth

As your website grows — whether you're blogging, selling products, teaching courses, or hosting videos — backend skills become essential. You'll use them to:

- Connect email marketing tools
- Add subscription or membership plans
- Analyze visitor data
- Back up your site for safety

The backend grows with you, but only if you know how to use it.

What You Can Do in the Backend (At a Glance)

Here's your high-level map of the WordPress dashboard — a quick overview of what each area does. Don't worry about mastering them all right now; you'll get plenty of practice later in the book.

- **Pages** – Permanent content like Home, About, Contact (see Chapter 4).
- **Posts** – Blog articles, updates, or news (see Chapter 5).
- **Media** – Your library for images, videos, and files.
- **Appearance** – Themes, menus, and widgets that shape your site's look.
- **Plugins** – Add features like e-commerce, SEO, or chatbots.
- **Users** – Manage who can log in and what roles they have.
- **Tools & Settings** – Import/export data, adjust site-wide preferences.
- **Security & Backups** – Protect your site from hackers and accidents.
- **E-commerce & Payments** – Manage products, orders, and gateways like Stripe or PayPal.

How This Chapter Fits Into Your Journey

Think of this chapter as your **map of the control room**. You don't need to memorize every detail right now. Instead, come back to this list as your website grows. Need to publish a blog? Flip to Posts. Selling products? Head to Plugins and E-commerce. Step by step, you'll learn the tools you need, when you need them.

Final Takeaway

Your backend is the **engine of your website**. The more you understand it, the faster and more confidently you can make updates, add features, and grow. Confidence doesn't come from avoiding the backend — it comes from exploring it.

In the next chapters, we'll break down these areas one at a time, so you can go from feeling overwhelmed to feeling like the pilot of your own website.

Chapter 7: Choosing (and Buying) a Domain Name for Your Online Store

When I first started building my website, one of the most exciting (and nerve-wracking) moments was picking the domain name. It felt like naming a business, a book, or even a child — something people would see first and remember me by.

I scribbled dozens of ideas in a notebook. Some were too long, some didn't sound professional, and some were already taken. Then, after trying a few combinations, I typed in **YourAIStudyBuddy.com**. To my surprise, it was available. Within minutes, it was mine — my little piece of the internet. That's when everything suddenly felt real.

So, what exactly is a domain name? Think of it this way:

- **Domain name** = the address people type (example: youraistudybuddy.com).
- **Website** = the house (your pages, images, and products).
- **Hosting** = the land the house sits on (where your files live).
- **Registrar** = the store where you buy and manage the address.
- **DNS/Nameservers** = the GPS system that points your address to the right house.

In simple terms: without a domain, people can't find your website.

Why Your Registrar Choice Matters

The company you buy your domain from (called a **registrar**) makes a big difference for beginners. It affects:

1. **Ease of setup** – Do they have a clear dashboard and simple buttons?
2. **Security** – Do they include WHOIS privacy, two-factor login, and domain lock?
3. **Reliability** – Does their DNS work smoothly so your site loads everywhere?
4. **Extras** – Can you easily add branded email addresses like *info@yourdomain.com*?
5. **Pricing** – Many companies give a low first-year price, then raise renewals.

💡 **Pro Tip:** Always check the renewal price, not just the promo price.

Registrar Must-Have Checklist ✅

Before you buy, make sure your registrar offers:

- Free **WHOIS privacy** (protects your personal info from spam).
- **Two-factor authentication (2FA)** for account security.
- A simple, beginner-friendly dashboard.
- Easy DNS editor and nameserver setup.
- Transparent renewal pricing.
- Responsive support (chat or ticket system).

How to Pick a Great Domain Name

Here are my rules of thumb:

1. **Keep it short and simple** – ideally 6–18 characters.
2. **Easy to say and spell** – pass the "radio test."
3. **Avoid hyphens and numbers** – they confuse people.
4. **Stick with .com** if possible — it's still the most memorable.
5. **Match your brand** – clear, trustworthy, and aligned with your mission.
6. **Future-proof it** – make sure it works as your business grows.
7. **Check social media handles** – consistency helps branding.

📝 **Mini Brainstorm Exercise:**

- Write 5–10 name ideas.
- Say each out loud.
- Ask 2 friends to spell it after hearing it once.
- Search availability with a registrar and shortlist 2–3.

Beginner-Friendly Registrars

Some registrars that are popular and trusted include:

- **Hostinger** – great beginner bundles (domain + hosting).
- **Namecheap** – clear pricing, free WHOIS privacy.
- **GoDaddy** – huge selection of extensions.
- **Cloudflare Registrar** – strong security, at-cost pricing.
- **Google Domains (now Squarespace)** – clean interface, great for Gmail integration.

Step-by-Step: Buying Your Domain

1. Go to a registrar (like Hostinger, Namecheap, or GoDaddy).
2. Search for your chosen name (yourbrand.com).
3. If it's taken, try small variations or other endings (.com, .ai, .shop).
4. Add to cart → choose 1 year to start (turn on auto-renew).
5. Turn on WHOIS privacy (if available).
6. Create your account and enable 2FA.
7. Pay and verify your email (required).

🎉 Congratulations — you now own your domain!

Connecting Your Domain to Your Website

If your hosting and domain are with the same company (like Hostinger), they'll usually connect automatically. Just follow their one-click setup for WordPress.

If your hosting is with a different provider:

1. Get your host's nameservers (they look like ns1.yourhost.com, ns2.yourhost.com).
2. Go to your registrar → Manage Domain → Nameservers.
3. Paste the new nameservers and save.
4. Wait for DNS to update (15 minutes–24 hours).

💡 **Pro Tip:** After changing nameservers, clear your browser cache or check your domain from your phone's data connection to see the updates sooner.

Common Beginner Mistakes 🚫 (and Quick Fixes)

1. **Buying only on price** – Renewal fees surprise you.
   → Fix: Always check renewal pricing first.
2. **Forgetting WHOIS privacy** – Spam starts rolling in.
   → Fix: Turn on privacy.
3. **Not enabling 2FA** – Makes your account easy to hack.
   → Fix: Always enable 2FA.
4. **Skipping SSL (https)** – Browser shows "Not secure."
   → Fix: Activate free SSL from your hosting provider.

Quick FAQ

Can I change my domain later?
Yes, but it's messy. You'll need redirects and may lose SEO. Choose carefully now.

Do I need .com?
Not always. .com is best, but .ai, .shop, or .org can work if they fit your brand.

How long does DNS take?
Sometimes minutes, but allow up to 24–48 hours for global updates.

End-of-Chapter Checklist ✅

- I picked 2–3 strong domain name options.
- I checked spelling, saying it out loud, and social handles.
- I chose a registrar with WHOIS privacy + 2FA.
- I bought the domain and verified my email.
- I connected my domain to my hosting.
- I activated SSL so my site is secure.
- (Optional) I set up a professional email like info@mydomain.com.

✨ **Final Takeaway:**
Buying a domain is more than just grabbing a name — it's staking your claim on the internet. Choose wisely, secure it properly, and you've just taken the first big step toward building your online presence.

Chapter 8: Establishing Your Brand Identity Through Your Domain Name

When I bought my first domain name, I had no idea how important it would become. To me, it was just the thing I needed to make my website live. But as I went through the process, I realized something: a domain name isn't just a technical step — it's part of your brand identity. It's how people find you, remember you, and talk about your business.

Think of it like opening a physical store:

- The **address (domain name)** tells people where to go.
- The **store sign (your brand)** tells them who you are.
- The **decor (your website design)** tells them your style and values.

If your domain name is confusing, hard to spell, or looks unprofessional, people may never even visit your site. That's how powerful it is.

1. What Exactly Is a Domain Name?

A domain name is simply the address people type into their browser to reach your website.

- Easy: **www.YourAIStudyBuddy.com**
- Hard: **192.0.2.123** (that's an actual IP address format — nobody is going to remember that).

Every domain has three parts:

1. **Prefix** – usually "www."
2. **Second-Level Domain (SLD)** – your unique name, like *youraistudybuddy.*
3. **Top-Level Domain (TLD)** – the ending like *.com, .net, .shop, .ca.*

When you put it together, you've got a one-of-a-kind identity online.

2. Why Your Domain Name Matters

Your domain is often the very first impression someone has of your business. If it feels trustworthy and professional, people are more likely to click. If it feels spammy or complicated, they'll skip it.

Bad Example: discountshoes247cheap.com → too long, looks sketchy, hard to remember.
Good Example: shoebox.com → short, clean, and easy to trust.

See the difference? One looks like a brand, the other looks like a scam.

3. How to Choose the Right Domain

Here's a process I wish I knew before buying my first one:

1. **Brainstorm names** – Write down 5–10 options. Use your brand name, or mix in keywords like "AI," "design," or "shop."
2. **Say it out loud** – Can someone spell it after hearing it once? That's called the "radio test."
3. **Check availability** – Use a registrar like Hostinger, Namecheap, or GoDaddy to see if your names are taken.
4. **Pick your extension (TLD):**
    - *.com* → most professional, global reach.
    - *.shop* → perfect for online stores.
    - Country codes (.ca, .uk, .au) → great for local trust.
5. **Avoid pitfalls:**
    - No hyphens.
    - No random numbers (unless it's part of your brand).
    - No trademarked names (to avoid legal trouble).

💡 **Pro Tip:** Future-proof your name. Pick something that still makes sense if you expand into new products or services later.

4. Buying Your Domain Name (Step by Step)

Let's use **Hostinger** as an example, since that's what I used for *YourAIStudyBuddy.com*.

1. Go to Hostinger → search your domain.
2. If it's available, click **Add to Cart.**
3. Choose your registration period (start with 1 year if you're unsure).
4. Turn on **WHOIS Privacy Protection** (keeps your info private).
5. Pay → confirm ownership via email.

🎉 Boom — you officially own your digital address.

5. Securing Your Domain

Buying it isn't enough — you need to lock it down, just like you would a house.

- **Enable SSL (https://)** → gives your site the padlock icon.
- **Turn on Domain Locking** → stops hackers from transferring it.
- **Activate WHOIS privacy** → hides your personal contact details.

6. Protecting Against Cybersquatters

Cybersquatters are people who buy domains similar to yours to resell them at a higher price — or worse, to mislead your customers.

How to protect yourself:

- Buy common versions (.net, .org, .shop).
- Register common misspellings.
- Keep your domain renewed so it doesn't expire.

Example: If your brand is **myflowerstore.com**, consider also grabbing:

- myflowerstore.net
- myflowerstore.shop
- myfloriststore.com

It's a small investment for peace of mind.

7. Going Global or Staying Local

The extension (TLD) you choose signals whether your site is meant for the world or just one country.

- **.com** → best for global businesses.
- **.ca, .co.uk, .au** → build trust locally.

For example: If you sell in Canada and the U.S., owning both **mybrand.com** and **mybrand.ca** shows professionalism.

8. Summary

Your domain name isn't just an address — it's part of your brand. A strong domain:

- Builds trust with visitors.
- Helps you rank better in search engines.
- Protects your business from copycats.
- Positions you for growth, whether local or global.

When I bought *YourAIStudyBuddy.com*, it felt like I had planted my flag on the internet. It wasn't just a technical step — it was me saying, *this is my space, this is my brand.* And now, you're about to do the same.

In the next chapter, we'll take the domain you've chosen and connect it to your website so your brand can officially go live.

Chapter 9: Installing WordPress on Your Computer – Your First Step

What Does It Mean?

Installing WordPress on your own computer means creating a **practice website** that only you can see. It's private, safe, and doesn't require buying a domain name or hosting yet. Think of it as a **sandbox** — a place to play, experiment, and learn before you launch your real site to the world.

Why This Matters for Beginners

When you're new to building websites, the idea of "breaking" something on a live site can feel overwhelming. Installing WordPress locally gives you:

1. Freedom to learn at your own pace.
2. A safe place to test themes, plugins, and layouts.
3. Cost savings (no hosting or domain yet).
4. Offline access (you can work without internet).

It's like learning to drive in a parking lot before heading onto the highway.

Step 1 – Understanding WordPress (Two Versions)

There are two types of WordPress:

1. **WordPress.com** – Hosted for you, but limited unless you pay. (Like renting an apartment: you can decorate, but you don't own it.)
2. **WordPress.org** – Free, open-source software you install yourself. (Like owning a house: you can remodel however you want.)

📌 **In this book, we use WordPress.org because it gives you full control.**

Step 2 – Why You Need a Local Server

Your computer doesn't naturally know how to run WordPress because WordPress requires:

- **PHP** (the programming language it's written in)
- **MySQL** (a database where your site's content lives)

A **local server** is a free program that makes your computer act like a web host so you can run WordPress privately.

Popular tools:

- **LocalWP** (best for beginners — our choice here)
- XAMPP
- MAMP

Step 3 – Downloading LocalWP

1. Go to localwp.com.
2. Click **Download for Free**.
3. Select your operating system (Windows or Mac).
4. Fill in your name/email (optional).
5. Download and run the installer.

📌 **Beginner Tip:** LocalWP is safe, free, and widely used by professionals.

Step 4 – Creating Your First Practice Site

1. Open LocalWP.
2. Click **+ Create a New Site**.
3. Enter a site name (e.g., *MyPracticeSite*).
4. Choose **Preferred Settings** when prompted.
5. Create a username and password (save them somewhere safe).
6. Click **Add Site**.

LocalWP will install WordPress for you behind the scenes.

Step 5 – Opening Your Practice Site

Once installed, you'll see two options:

- **Open Site** – See your website like a visitor would.
- **WP Admin** – Log in to your private dashboard.

Click **WP Admin**, enter your username/password, and congratulations — you're inside WordPress! 🎉

Step 6 – Safe Practice Zone

Here's what you can try without risk:

- Add and format pages.
- Install and test plugins.
- Write practice blog posts.
- Experiment with themes and colors.

No one else can see your site yet. This is your **learning lab.**

Common Mistakes 🚫

1. Forgetting your login details → Write them down or use a password manager.
2. Expecting others to see your local site → Local sites are private until you move them online.
3. Skipping backups → Even practice sites should be backed up occasionally.

Quick Checklist ✅

- Installed LocalWP successfully.
- Created a practice site with username/password saved.
- Logged into WordPress Admin (backend).
- Tested adding a page or post.

Final Takeaway

By installing WordPress on your computer with LocalWP, you've created a safe space to practice. This is your training ground before launching a live site.

Later in this book, we'll take everything you learn here and move it onto real hosting (like Hostinger), so your website can go live for the world to see.

Chapter 10: Installing WordPress — Two Ways to Start Your Website

Before you can design a homepage, add products, or connect your chatbot, you first need WordPress installed. Think of this step as laying the foundation before building a house — without it, there's nothing to build on.

In this chapter, we'll cover the **two main ways** to install WordPress so you can choose the one that fits your goals.

Why You Need to Install WordPress First

WordPress is the **engine** that makes your website run. Without it, your domain is just an empty lot on the internet. Installing WordPress gives you the tools to:

- Create and edit pages
- Write blog posts

- Install plugins (like e-commerce or chatbots)
- Design and manage your website

Option 1: Install WordPress Through Your Hosting (Live Website) ✅ Recommended for This Book

This is the fastest and easiest way to get your website online. It's ideal if you want to launch your site right away and follow along step-by-step in this book.

How It Works

Most hosting companies (like Hostinger, Bluehost, SiteGround) offer a **one-click installer** that sets up WordPress on your domain.

Step-by-Step (using Hostinger as an example):

1. Log in to your Hostinger account.
2. In the menu, click **Website → Auto Installer**.
3. Select **WordPress** from the list.
4. Enter your site title, email, username, and password (write these down!).
5. Choose the domain you want to connect.
6. Click **Install**.

In a minute or two, you'll get two important links:

- **View Site** → what visitors see
- **Admin Dashboard** → your private control panel

💡 **Beginner Tip:** This is the method we'll use throughout the rest of the book because it sets up a *real, live website* you can share with the world.

Option 2: Install WordPress on Your Computer (Local Site) 🖥 Practice Mode

This option is for practicing in private before going live. It's like a sandbox — you can test things without the world seeing.

To do this, you'll need software that turns your computer into a "fake" web server.

Popular Tools:

- **LocalWP** (beginner-friendly — our top pick)
- XAMPP
- MAMP

Step-by-Step (using LocalWP):

1. Go to localwp.com and click **Download**.
2. Choose your operating system (Windows or Mac).

3. Install and open the program.
4. Click **Create New Site**.
5. Give your site a name (e.g., *PracticeStore*).
6. LocalWP installs WordPress automatically.
7. Click **WP Admin** to log in and start exploring.

💡 **Beginner Tip:** A local site is only visible to you. When you're ready to go live, you'll still need to install WordPress on your hosting (Option 1).

Which Should You Choose?

- **Go with Option 1 (Hosting install)** if you're ready to launch your store now.
- **Go with Option 2 (Local install)** if you want a safe place to practice first.

For this book, we'll focus on **Option 1** so you can build your real site step-by-step. But it's good to know both methods for future projects.

Beginner Mistakes to Avoid

1. **Forgetting your login info** → Write down your username, password, and login link.
2. **Installing twice by accident** → Causes conflicts. Install once, confirm it works.
3. **Choosing the wrong domain** → Double-check before you hit "Install."
4. **Skipping SSL setup** → Always make sure your site uses **https://** for security.

Summary

Installing WordPress is your first real step toward building an online presence. Whether you choose the **live site route** (via hosting) or the **practice route** (on your computer), you now have the foundation in place.

👉 In the next chapter, we'll take a **guided tour of your WordPress Dashboard** so you know exactly where everything lives and how to start shaping your website.

Chapter 11: Touring Your WordPress Dashboard – Your Website's Control Center

Congratulations! 🎉 If you've followed along, you now have WordPress installed and a website foundation ready to build on. When you log in for the first time, you'll land in the **WordPress Dashboard**.

Think of the Dashboard as the **control room** of your website. This is where you'll create pages, write blog posts, upload images, install plugins, manage settings, and more. Everything starts here.

First Look: What You See in the Dashboard

When you log in, the Dashboard may look a little overwhelming at first, but don't worry — it will quickly start to feel familiar. On your screen, you'll see:

1. **Top Bar (Admin Bar)** – runs across the top. Gives quick access to your site, notifications, and shortcuts.
2. **Left Menu (Sidebar)** – the main navigation for everything you'll do.
3. **Main Area (Workspace)** – changes depending on what menu item you click.
4. **Welcome Panel** – quick links to help you start (you can dismiss this later).

💡 **Beginner Tip:** If you ever feel lost, look to the left menu. It's your "map."

The Left Menu: Your Key Sections

Let's break down the most important menu items, one by one.

1. **Dashboard**
   - *What it does:* Quick overview of your site (updates, comments, site health).
   - *Why it matters:* Shows you what needs attention right away.
2. **Posts**
   - *What it does:* Create and manage blog articles.
   - *Why it matters:* Posts are how you keep your site fresh and attract visitors.
3. **Media**
   - *What it does:* Stores all your images, videos, PDFs, and downloads.
   - *Why it matters:* Keeps your visuals organized.
4. **Pages**
   - *What it does:* Build permanent content like your Home, About, and Contact pages.
   - *Why it matters:* These are the backbone of your site.
5. **Comments**
   - *What it does:* Manage feedback from visitors.
   - *Why it matters:* Keeps you connected with your audience.
6. **Appearance**
   - *What it does:* Choose your theme, customize colors, menus, and widgets.
   - *Why it matters:* This is where your brand's look and feel comes alive.
7. **Plugins**
   - *What it does:* Add extra features (SEO tools, e-commerce, chatbots).
   - *Why it matters:* Transforms your site from basic to powerful.
8. **Users**
   - *What it does:* Manage who can log in and what they can do.
   - *Why it matters:* Useful if you have a team or guest writers.
9. **Tools**
   - *What it does:* Import/export content, run plugin tools.
   - *Why it matters:* Advanced, but handy as your site grows.
10. **Settings**
    - *What it does:* Manage site title, URL, language, time zone, permalinks.
    - *Why it matters:* Critical for making your site run correctly.

💡 **Beginner Tip:** Don't worry if you don't understand everything yet. You'll use different sections as needed — and this book will guide you step by step.

Dashboard Pro Tips for Beginners

- **Collapse the menu** – Click the small arrow at the bottom of the sidebar to give yourself more space.
- **Screen Options tab** – On some pages, you can choose what info shows up.
- **Updates tab** – Keep WordPress, themes, and plugins updated for security.
- **Help tab** – At the top of most pages, you'll find extra tips.

Avoid These Common Beginner Mistakes 🚫

1. Ignoring updates → leads to security risks.
2. Installing too many plugins → slows your site.
3. Changing settings without knowing → can break links or layouts.
4. Forgetting to write down your login info.

Summary

The WordPress Dashboard is your **control center**. At first, it may feel like a cockpit full of buttons — but as you explore, you'll learn what each one does. With this foundation, you'll feel confident moving forward.

👉 In the next chapter, we'll start customizing your site's **Appearance and Theme** — the step that makes your website look unique and aligned with your brand.

Chapter 12: Exploring the WordPress Dashboard – Your Website's Control Center

Now that WordPress is installed, it's time to step inside the **Dashboard** — the control center of your entire website. Think of it as the cockpit of an airplane: at first it looks intimidating, but once you learn what each button does, you'll feel confident taking off.

This chapter gives you a guided tour of the dashboard and reveals the "hidden layers" behind the tasks you'll use most often.

Why the Dashboard Matters

Your dashboard is where you:

- Create and edit **pages** (Home, Products, Terms & Conditions, etc.)
- Write **blog posts** to attract and engage readers
- Upload and manage **images and media**
- Approve or moderate **comments**
- Install and configure **plugins and themes**
- Connect tools like **email marketing services** or **payment gateways**
- Set up **menus** for navigation
- Secure your site with **backups and firewalls**

It's more than just a menu — every decision you make here shapes how your site looks, works, and grows.

Pages – Building Blocks of Your Site

On the surface, creating a page seems simple: click "Add New Page." But behind the scenes, you'll need to:

1. Choose a **layout** (block editor or classic editor).
2. Add content — text, images, or buttons.
3. Set a **featured image**.
4. Decide whether it's a **parent page** or a **child page** (part of a section).
5. Add it to your **navigation menu**.
6. Choose visibility (public, private, or password protected).
7. Add SEO elements (title tags and meta descriptions).

👉 It's not just about putting words online — it's about structure, user experience, and how Google finds your page.

Posts – Dynamic Content That Builds Authority

Writing a post is more than typing a few paragraphs. Every post involves:

- Choosing **categories and tags** to organize your content.
- Structuring with headers (H1, H2, H3) for readability.
- Embedding **images or videos**.
- Setting a **publish date** (or scheduling for later).
- Selecting **SEO keywords**.
- Deciding whether comments are allowed.
- Writing an **excerpt** for previews.
- Linking to other posts or external resources.

👉 A blog post is a tool for engagement, SEO, and authority — not just a "journal entry."

Media – More Than Just Uploading Images

When you upload an image, you're also making choices that affect speed, SEO, and accessibility. Best practices include:

1. Resizing and compressing images before upload.
2. Naming files clearly (e.g., `ai-guide-cover.jpg`, not `IMG001.jpg`).
3. Writing **alt text** for accessibility and SEO.
4. Choosing placement and alignment.
5. Using featured images for posts/pages.
6. Creating galleries or sliders if needed.

👉 Unoptimized images are one of the biggest reasons websites load slowly.

Comments – Managing Community Engagement

Comments can build community — or drown your site in spam. Behind every comment, you'll need to:

- Install an anti-spam plugin like **Akismet**.
- Approve or trash comments.
- Respond to readers to encourage discussion.
- Decide whether comments are allowed everywhere or only on blog posts.
- Set notifications so you don't miss new feedback.

☞ Good comment management boosts trust; neglecting it can hurt your brand.

Plugins & Themes – Expanding Your Website's Powers

A plugin is like an **app for your website**. Themes define your site's design. But it's not as simple as hitting "Install." You must:

- Decide between free vs. premium.
- Check reviews and update history.
- Test compatibility with your WordPress version.
- Avoid installing too many (they can slow or break your site).
- Learn each plugin's setup options.

☞ Think of plugins as tools in your toolbox — choose wisely, and your site becomes powerful.

Email Tools – Connecting With Visitors

Setting up email is critical for marketing, but it comes with hidden steps:

1. Choose a service (MailerLite, Mailchimp, ConvertKit, etc.).
2. Connect your forms to that service.
3. Enable **double opt-in** for legal compliance.
4. Create welcome emails or sequences.
5. Add SPF/DKIM records to prevent delivery issues.

☞ Done right, email builds long-term relationships. Skipped steps = lost subscribers.

Payments – Adding Stripe or PayPal

To accept payments, you'll need to:

1. Sign up for Stripe or PayPal.
2. Install the correct plugin (e.g., WooCommerce Stripe Gateway).
3. Secure your site with **SSL** (https://).
4. Create products or services in your dashboard.
5. Test checkout using test card numbers.
6. Set up refunds and cancellations.

👉 Payments must be airtight. One mistake can cost you sales.

Menus – The Roadmap for Visitors

Menus control how people move through your site. Setup involves:

- Choosing menu locations (header, footer, sidebar).
- Deciding what pages/posts/categories to include.
- Reordering with drag-and-drop.
- Nesting pages for dropdowns.
- Creating mobile-friendly versions.

👉 A confusing menu drives visitors away — a clear one keeps them exploring.

Security – Protecting Your Work

Every website needs protection. You'll want to:

1. Install a security plugin (Solid Security or Wordfence).
2. Enable **two-factor authentication**.
3. Change the default login URL.
4. Limit login attempts.
5. Schedule **backups**.
6. Keep plugins and themes updated.
7. Run regular malware scans.

👉 Security isn't automatic — it's an ongoing responsibility.

Final Takeaway

Your **WordPress Dashboard** is much more than a menu of buttons. It's where your site's content, design, marketing, payments, and security all come together.

At first, it may feel overwhelming. But the more you explore, the more confident you'll become — and soon, managing your site will feel second nature.

In the next chapter, we'll dive into **WooCommerce** — the plugin that transforms your WordPress site into a fully functional online store.

Chapter 13: The WordPress Dashboard Demystified

Your Website's Control Panel

When you log in to WordPress, the first thing you'll see is the **Dashboard** — your website's control panel.

Think of it like the **cockpit of an airplane**: every button, switch, and gauge controls some part of your site's performance, design, or content. At first, it looks busy, but once you know what's what, you'll feel confident piloting your website.

This chapter is your **map to each dashboard section**, so you'll always know where to go and what it does.

Why the Dashboard Matters

- It's the **central hub** for managing content, design, and settings.
- You can access **every feature from one place** — no endless searching.
- The more familiar you are with it, the faster and more confidently you can make changes.

Main Sections of the WordPress Dashboard

1. Dashboard

- **What it does:** Shows recent activity, plugin updates, and site health.
- **Why it matters:** A quick snapshot of what needs your attention.

2. Posts

- **What it does:** Stores and organizes your blog articles or news updates.
- **Why it matters:** Lets you add, edit, and categorize time-stamped content. (See Chapter 5 for details.)

3. Media

- **What it does:** Your library of images, videos, PDFs, and files.
- **Why it matters:** Keeps all your visual assets organized for easy reuse.

4. Pages

- **What it does:** Houses permanent content like Home, About, and Contact.
- **Why it matters:** Forms your site's core structure. (See Chapter 4 for details.)

5. Comments

- **What it does:** Displays visitor feedback on posts or products.
- **Why it matters:** Lets you approve, reply, delete, or mark comments as spam.

6. Appearance

- **What it does:** Controls your site's look and layout (themes, menus, widgets).
- **Why it matters:** Shapes your brand's visual identity.

7. Plugins

- **What it does:** Adds extra features like e-commerce, SEO tools, or chatbots.
- **Why it matters:** Extends your site's capabilities without coding.

8. Users

- **What it does:** Manages who can log in and what they can do (Admin, Editor, Subscriber).
- **Why it matters:** Keeps your site secure with proper roles and permissions.

9. Tools

- **What it does:** Offers import/export options, site health checks, and plugin-specific tools.
- **Why it matters:** Useful for migrations, backups, and troubleshooting.

10. Settings

- **What it does:** Adjusts your site's title, URL, language, and global preferences.
- **Why it matters:** Ensures your site runs smoothly and is SEO-ready.

Extra Menu Items (After Installing Plugins)

As you add plugins, new menu sections may appear, such as:

- **WooCommerce** – Manage products, orders, and payments.
- **Products** – Add and edit store items.
- **WP Mail SMTP** – Configure email delivery.
- **Botpress Chat** – Manage your chatbot.
- **Stripe** – Set up your payment gateway.

Quick Tips for Navigating the Dashboard

- Collapse the menu (small arrow at the bottom) for more workspace.
- Use the **search bar** (in newer WordPress versions) to find menu items quickly.
- Regularly check **Dashboard → Updates** to keep your site secure.

Final Thoughts

Your WordPress Dashboard doesn't need to feel intimidating.

Once you know what each section does and where to find it, you'll **navigate like a pro** — spending less time searching for settings and more time creating, customizing, and growing your site.

Now that you understand the control panel, you're ready to dive into the features that power your website's growth.

Chapter 14: Installation & Hosting

How I Got My WordPress Site Up and Running From Scratch

Starting my website wasn't about having all the answers — it was about taking the first step. And for me, that first step was learning how to **host a website** and install **WordPress the right way**.

If you're feeling overwhelmed or unsure where to begin, don't worry — I've been there. In this chapter, I'll walk you through exactly what I learned, one step at a time.

Before I could design pages, upload content, or connect my chatbot, I had to figure out the foundation:

1. What is hosting?
2. How do I install WordPress?
3. What's the difference between themes and plugins?

At first, these questions felt technical. But once I understood them, it was like snapping puzzle pieces into place — and the end result was something I was proud of: **YourAIStudyBuddy.com**.

Understanding What Hosting Really Means

Let's simplify this: hosting is like the **land your website is built on**. Just as a physical store needs a location, your digital store or blog needs space on the internet.

A web host provides that space by storing your site's files on servers and making them accessible 24/7. Without hosting, your website has nowhere to live. You might have a beautiful WordPress design, but no one will see it unless it's uploaded to a server.

That's why choosing the right hosting provider is one of the most important first steps.

Choosing the Right Hosting Provider

When I began, I didn't know one hosting company from another. But after research and talking to other entrepreneurs, I realized hosting is not just about **price** — it's about **peace of mind**.

Here's what I looked for:

1. ☑ **Speed & Performance** – Fast sites keep visitors happy and rank higher on Google.
2. ☑ **Uptime Reliability** – I wanted at least 99.9% uptime so my site wouldn't randomly disappear.
3. ☑ **Customer Support** – As a beginner, I needed 24/7 help when things got confusing.
4. ☑ **Ease of Use** – One-click WordPress installation was non-negotiable.
5. ☑ **Affordability** – Quality mattered, but I also had a startup budget.

After weighing my options, I chose **Hostinger** because it checked every box: beginner-friendly, secure, reliable, and affordable.

Other good beginner-friendly providers include **Bluehost, SiteGround,** and **GoDaddy** — each has its strengths depending on your needs.

Installing WordPress the Simple Way

Most modern hosts let you install WordPress with just **one click** — and that's exactly what I did. Inside my Hostinger dashboard, I found the **Auto Installer**, selected WordPress, and entered a few details:

1. My website name
2. My admin username and password
3. My email address

Within minutes, WordPress was installed. Logging into my Dashboard for the first time felt like my vision was finally coming to life.

📌 **Beginner Tip:** If your hosting provider is different, don't worry — most offer similar tools. And if you get stuck, support teams can walk you through it in minutes.

What Are Themes and Plugins?

Once WordPress was installed, my site was technically live — but it didn't look like much yet. That's because WordPress starts as a **blank canvas**. Themes and plugins are what bring it to life.

✅ Themes — Your Site's Look and Feel

Themes control your site's **design**: colors, fonts, layout, and overall style. I chose a clean, modern theme that felt approachable and professional. WordPress offers thousands of free and premium themes to match your brand.

✅ Plugins — Your Site's Tools and Features

Plugins are like **apps** for your website. They add features WordPress doesn't include by default. For **YourAIStudyBuddy.com**, I installed:

- **Stripe Payment Gateway** – For secure online payments
- **Download Manager** – To share PDFs and worksheets
- **Botpress Chatbot** – To guide users and answer questions
- **Antispam & Security Plugins** – To keep my site safe
- **SEO Tools** – To improve visibility on Google

Installing a plugin is simple:

- Go to **Plugins → Add New** in your Dashboard
- Search for the plugin
- Click **Install**, then **Activate**

Looking Back — and Looking Forward

Building a website felt overwhelming at first. But once I understood hosting, learned how to install WordPress, and discovered how themes and plugins worked, it all began to **click**.

WordPress.org gave me the freedom to create a website that reflects my vision and mission — without hiring a developer. It proved that even without a tech background, I could build something powerful, professional, and profitable.

In the next chapter, I'll share how I connected my **domain name**, structured my site, and began uploading real content — step by step, so it never feels overwhelming.

Because if I can do it, **you absolutely can too**.

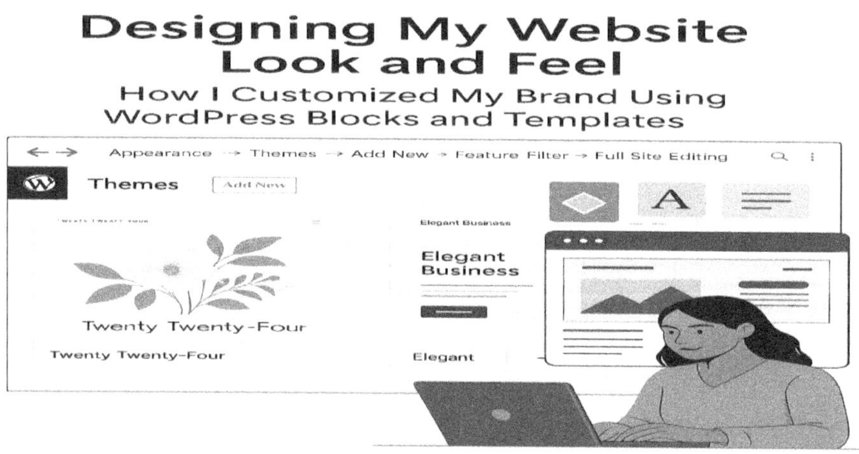

Chapter 15: Choosing and Installing the Perfect WordPress Theme

1. What Is a Theme (and Why It Matters)?

- A theme = your website's **outfit**. Controls design, layout, colors, fonts.
- You can swap designs without losing content.
- Demo vs reality: your site may not look like the preview until you customize.

2. The Four Types of WordPress Themes

- **Block Themes** (future of WP, full site editing, flexible, no code).
- **Classic Themes** (older, use Customizer + widgets, huge library).
- **Hybrid Themes** (classic base but partial block editing).
- **Universal Themes** (switch between both systems).
  👉 Beginner Tip: If you're starting fresh today, choose a **block theme**.

3. How to Choose the Right Theme

- Use **Appearance → Themes → Add New → Filters**.
- Check: Popular, Latest, Block Themes.
- Evaluate reliability: update history, reviews, support, active installs.
- Free vs Paid: Free = great to start, Paid = premium features + support.

4. How to Install a Theme

Method 1: Repository → Appearance → Themes → Add New → Install → Activate.
Method 2: Upload Zip → Add New → Upload → Install → Activate.

5. How to Manage Themes

- **Update**: Dashboard → Updates (always back up first, use child themes).
- **Delete**: Appearance → Themes → Select Theme → Delete.

6. Customizing Your Theme with the Site Editor

(Only for block themes)

- **Header:** Add logo, menus, tagline.
- **Footer:** Add address, hours, social icons.
- **Homepage:** Bold welcome message + CTA.
- **Styles Panel:** Set colors, fonts, buttons, consistency.
- **Template Parts:** Reuse headers/footers across pages.
- Safe to experiment → Preview + Revisions built-in.

☆ Case Study: My Journey with Block Themes

- Why I chose a block theme for my eCommerce site.
- How I customized my shop layout, checkout, and blog section.
- Tools I used: Site Editor, Styles Panel, Templates, Template Parts.
- The big difference I noticed vs Classic Themes.

7. Getting Help and Support

- Use the WordPress.org theme directory → Support Forum.
- Search for solutions or ask theme developers directly.

Final Thoughts

Choosing a theme is exciting — it's where your website starts to look like *your* brand. Don't stress about picking the "perfect" one immediately. The beauty of WordPress is you can experiment, preview, and change anytime without losing your content.

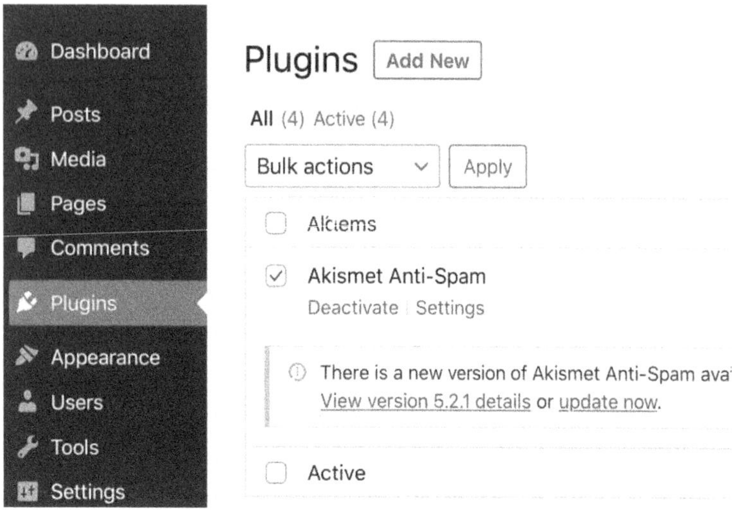

Chapter 16: WordPress Plugins — Apps for Your Website

Plugins are what make WordPress so powerful. Think of them like apps on your smartphone: you install only the ones you need to extend your site's features — no coding required.

With plugins, you can:

- Accept payments (WooCommerce, Stripe)
- Create contact forms (WPForms)
- Improve site speed (LiteSpeed Cache)
- Boost your SEO (Yoast or Rank Math)
- Add galleries, social sharing buttons, or even an AI chatbot

There are over **60,000 plugins** in the official WordPress directory. Many are free, with optional paid upgrades for advanced features.

Why Plugins Matter

Plugins let you build exactly what you need — from a simple blog to a full online store — without hiring a developer. Just be careful not to overload your site; too many plugins can slow it down or cause conflicts.

Installing a Plugin

From your WordPress dashboard:

1. Go to **Plugins → Add New**
2. Search for what you need
3. Click **Install Now**, then **Activate**

That's it. Most plugins come with their own setup instructions.

My First Plugins

When I built my eCommerce site, I started small:

- **WooCommerce** — for products and payments
- **WPForms** — for contact forms
- **LiteSpeed Cache** — for speed
- **Solid Security** — for protection
- **SEO plugin** — to get found on Google

Each one served a specific purpose and transformed my site from a blank page into a working business.

Final Tip

Keep it simple. Add only the plugins you really need, update them regularly, and delete the ones you're not using. With the right selection, plugins turn WordPress into a powerful platform for anything you want to build.

Chapter 17: Securing My Website

How I Protected My Store from Hackers, Spam, and Downtime — No Tech Background Needed

When I first set up my online store, I thought my biggest challenge would be getting customers. Instead, I quickly realized that keeping my site **safe** was just as important. This wasn't just a blog anymore — it was a business. People would be trusting me with their money and information, and I had to earn that trust by protecting it.

At first, website security sounded intimidating. I didn't know what "brute force" meant, and the only "bots" I knew were from sci-fi movies. But I told myself: *If I can build a site, I can learn to secure it.* Step by step, I did — and you can too.

Choosing a Secure Host

Security begins before you even log in. Your host is like your landlord — if the building isn't safe, your store isn't either.
I chose Hostinger because it included:

- **Free SSL Certificate** – Adds the "https://" lock in browsers for secure transactions.
- **Daily Backups** – Automatic safety net.
- **DDoS Protection** – Shields against traffic overload attacks.
- **24/7 Monitoring** – Alerts if something looks suspicious.

👉 **Pro Tip:** If your host doesn't offer these, you'll need to add them yourself — don't skip this step.

Adding a Security Plugin

Inside WordPress, I installed **Solid Security** (free version). It added:

- **Brute Force Protection** – Stops hackers from guessing passwords.
- **File Change Detection** – Warns if sneaky files appear.
- **Two-Factor Authentication** – Requires a code from my phone to log in.
- **Database Backups** – Safeguards all customer and order data.

It felt like adding locks, cameras, and alarms to my store.

Strong Passwords & User Roles

I retired the "flower123" password forever. Now, my logins are long, random, and unique.
I also learned to assign roles wisely: a helper uploading products only gets "Shop Manager," not "Admin." That way, even if their account is hacked, damage is limited.

Blocking Spam & Bots

Spam hit me early — fake comments and bogus forms. My fixes:

- **Akismet Anti-Spam** – Blocks junk comments.
- **Google reCAPTCHA v3** – Invisible bot protection on forms and checkout.

Result? Customers shop smoothly, bots get blocked.

HTTPS & Backups

Switching to **https://** wasn't optional — it's required for secure payments. My host's free SSL made it easy.

For backups, I doubled up:

- **Hostinger's daily backups**
- **UpdraftPlus weekly backups to Google Drive**

If disaster strikes, I can restore my site in minutes.

Staying Updated

Hackers love old software. I turned on auto-updates for WordPress, themes, and plugins under:
Dashboard → Updates → Enable Auto-Updates

Monitoring Activity

Solid Security showed me real-time login attempts and changes. Once, I spotted multiple failed logins from overseas — and blocked them immediately.

Privacy & Trust

Since my store collects customer info, I created a clear **Privacy Policy** page. Not just for legal reasons, but to reassure visitors: *Your trust matters to me.*

Lessons Learned

- Security isn't optional — it's part of running an online store.
- You don't need to be a tech expert, just consistent.
- The right tools and small daily habits protect your brand and your customers.

Now, when I see that little lock icon next to my website name, I know it's more than a symbol — it's proof of the work I put in to keep my business safe.

Chapter 18: Getting Organized — Creating Pages, Menus, and a Smooth Learning Journey

Building a website isn't just about putting content online — it's about organizing it so visitors can easily find what they need. With *YourAIStudyBuddy.com*, I wanted every parent, student, and teacher to feel confident as they clicked through, never wondering "Where do I go next?"

That meant building a clear page structure, creating logical menus, and designing a learning journey that flowed naturally.

Step 1: Planning the Pages

I began with a simple list of the essentials:

Core Pages

- Home
- About
- Contact
- FAQ

- My Account

Learning & Content Pages

- Learn AI
- Watch Videos
- Read the Dictionary
- Books & Resources
- Meet Your AI Buddy (for kids)

Business & Conversion Pages

- Choose Your Plan
- Subscribe Now

Utility Pages

- Registration
- Login / Lost Password
- Payment Confirmation / Payment Failed

👉 **Pro Tip:** Keep page names clear and friendly. Visitors should know instantly what they'll find.

Step 2: Building Pages in WordPress

From the dashboard, I went to:
Pages → Add New

I used the block editor to add text, images, and layouts. At first, I kept things simple — the goal was to get the content live. Later, I added polish:

- Flipbook previews of books
- Call-to-action buttons
- Embedded videos
- Subscription grids
- Testimonials from parents and teachers

Each page became its own experience.

Step 3: Creating the Menu

A site is only as good as its navigation. From **Appearance → Menus**, I created:

Main (Top) Menu

- Home
- Learn AI
- Videos

- Books
- Choose Your Plan
- My Account
- Contact

Footer Menu

- Terms & Conditions
- Privacy Policy
- Support
- Login / Register
- Lost Password
- Social Media links

👉 **Pro Tip:** Keep menus clean (5–7 items). On mobile, test every link — small screens make clutter frustrating.

Step 4: Connecting the Journey

I didn't just make pages — I thought like a visitor:

- From Home → "Learn AI" button leads to the first video.
- From a video → a pop-up suggests checking the Dictionary.
- From the Dictionary → links guide visitors to the Subscription Plans.
- After subscribing → full access unlocks instantly (library + chatbot).

Like chapters in a book, the site flows step by step.

Step 5: Adding Trust & Personality

I wanted visitors to feel this was personal, not corporate. So I added:

- Photos of Raquel and her AI Buddy
- My story as a grandmother & entrepreneur
- A welcome message from me: *"Susie Hala, Creator of IntelliGloss"*
- Real images, not stock photos

Trust builds loyalty. These touches make the site feel alive and authentic.

Step 6: Testing and Tweaking

Finally, I tested the site on both laptop and phone — clicking every button, filling every form, even running a full checkout. Along the way I found and fixed:

- Broken links
- Headings too small on mobile
- A missing Stripe checkout button
- A video that didn't load properly

👉 **Pro Tip:** Always test your site like a customer would. What seems obvious to you might not be obvious to them.

Final Thoughts

Organization is more than menus and buttons — it's about guiding people on a journey that feels natural, trustworthy, and even enjoyable. When your site flows smoothly, visitors stay longer, explore deeper, and are more likely to become paying customers.

Chapter 19: Adding Stripe, Securing Payments, and Launching with Confidence

No eCommerce site is complete without a payment gateway. After building my pages, menus, and a smooth user experience, it was time to add one of the most important pieces: a secure, reliable checkout system.

I chose **Stripe** — not just because it's widely trusted, but because it's easy to integrate with WordPress, works for both subscriptions and one-time purchases, and gives customers a checkout experience that feels professional and safe.

Step 1: Why Stripe Was the Right Fit

Stripe offered exactly what I needed:

1. Supports both subscriptions and one-time payments.
2. Integrates smoothly with WordPress and WooCommerce.
3. Provides a fast, professional checkout experience.
4. Comes with a clean dashboard to manage transactions.

Most importantly, parents, students, and teachers would recognize Stripe's name and feel secure using it.

Step 2: Connecting Stripe to WordPress

After creating my Stripe account, I connected it to WordPress:

From the dashboard:
Plugins → Add New → WP Simple Pay (or WooCommerce Stripe Gateway)

Then I:

1. Copied my Test and Live API Keys from Stripe.
2. Pasted them into the plugin's settings.
3. Tested payments in **Test Mode** first.
4. Switched to **Live Mode** when everything worked.
5. Linked my success and failure pages (e.g., Payment Confirmation, Payment Failed).

👉 **Pro Tip:** Make sure your site is running on **https://** (SSL certificate) before accepting payments.

Step 3: Creating My Pricing Plans

On *YourAIStudyBuddy.com*, I launched with four subscription levels:

Plan	Price	Features
⚪ Free Trial	30 days	Basic video access, preview of AI Dictionary
⚫ Standard	$10/mo	Full video access, worksheets, Dictionary
⚫ Premium	$19.99/mo	All of Standard + 1-on-1 coaching
◯ Elite	$29.99/mo	All of Premium + custom learning paths + chatbot

I built a pricing table with buttons, each linking directly to the correct Stripe checkout page.

Step 4: Testing the Payment Flow

I pretended to be a real customer:

- Selected a plan
- Entered card details in **Test Mode**
- Completed checkout
- Received email confirmation + Stripe receipt

Once everything worked smoothly, I switched to Live Mode and published the real pricing.

Step 5: Building Trust With Customers

To make customers confident about checking out, I added:

- A **"100% Secure Checkout"** badge
- Stripe-powered payment logo
- FAQ section ("What happens after I subscribe?")
- Links to Terms & Conditions + Privacy Policy

This combination gave my store a polished, trustworthy look.

Step 6: The First Sale

I'll never forget the moment I got my first real notification from Stripe:
"You just made a sale."

It wasn't just money—it was proof. Proof that someone believed in my idea, trusted my site, and wanted to start learning with me. That moment made all the late nights building my store worthwhile.

And the best part? It was only the beginning.

Chapter 20: Adding a Chatbot to Your WordPress Site – My Step-by-Step Botpress Journey

When I started building *YourAIStudyBuddy.com*, I wanted it to be more than a library of books and videos. I wanted it to feel interactive—like a real study buddy was there to guide students and answer questions. That's when I discovered **Botpress**, an open-source chatbot builder powered by AI.

Why I Chose Botpress

After testing other chatbot tools like Tidio and ManyChat, I chose Botpress because:
✅ It uses natural language understanding (NLU).
✅ I could train it with my own content.
✅ It integrates smoothly with WordPress.
✅ It's open source and affordable.

It felt like hiring a 24/7 tutor—except this one never sleeps.

Building My First Chatbot

1. **Signing Up** – I created a free Botpress account and named my workspace *Your AI Study Buddy*.
2. **Designing Flows** – Inside Botpress Studio, I built simple conversation flows:
   - A welcome message ("Hi there! 👋 I'm your AI Study Buddy…")
   - FAQs like *"What is AI?"* or *"Where can I buy your book?"*
   - Buttons for quick replies
   - Fallback responses when the bot didn't understand
3. **Personality & Tone** – I gave the bot a friendly, supportive voice, like a tutor who's always there to help.

Embedding the Chatbot in WordPress

To make my bot live on my site:

1. I copied the embed code from Botpress → *Deploy* → *Webchat* → *Embed Code.*

2. In WordPress, I opened **Appearance → Widgets** and pasted it into a **Custom HTML block** in the footer.
3. (Alternative: Use the *Insert Headers and Footers* plugin if you don't want to edit theme files.)
4. Saved changes → refreshed my site → and there it was! My chatbot popped up, ready to chat.

Training and Customizing

I trained the bot with my own content so it could:

- Explain AI terms in simple language.
- Recommend books and videos.
- Guide users to subscription plans.
- Detect plan levels (Free, Standard, Premium, Elite) and upsell politely.

I also matched the design to my brand—soft colors, friendly messages, and a student-focused tone.

Testing With Real People

Before going live, I asked family, friends, and teachers to try it. Their feedback:

- "It feels like a real tutor!"
- "It's easier to find what I need now."
- "This bot is actually helpful—not just a gimmick."

That gave me confidence I had built something meaningful.

Lessons Learned

- Be patient—the first version won't be perfect.
- Train the bot with real user questions.
- Keep refining. A chatbot is always learning.

Final Thoughts

Adding a chatbot wasn't just a tech upgrade—it was a personal achievement. Now, when someone lands on *YourAIStudyBuddy.com*, they're greeted by a friendly AI assistant who is smart, helpful, and always available. And the best part? **I built it. And so can you.**

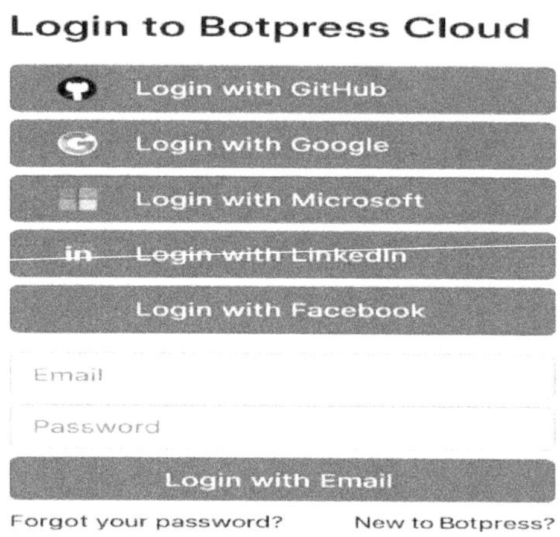

Chapter 21: Connecting the Chatbot – Turning Your Site Into an AI Buddy

In Chapter 20, I built and launched my first chatbot. But adding a chatbot was only the beginning. What made the difference was connecting it to my content, subscription system, and learning flow—so it became more than just a bot. It became an **AI Study Buddy**, guiding visitors step by step through their learning journey.

Step 1: From Chatbot to AI Buddy

At first, the chatbot could answer FAQs and greet users. That was helpful, but I wanted more. I wanted it to act like a tutor:

- Suggesting the next video after a student watched one
- Guiding parents to subscription plans based on their needs
- Explaining AI terms in simple language when students got stuck

This shift—from answering questions to **teaching and guiding**—is what transformed my site.

Step 2: Connecting to Content

I trained the chatbot with my own resources:

- Books and flipbooks
- Video lessons
- The AI Dictionary (IntelliGloss)
- Subscription plan details

Now, instead of sending users to random pages, the bot could **recommend specific resources**. For example:

- "Want to learn what AI is? Here's our starter video."
- "Looking for a kid-friendly explanation? Try the AI Buddy dictionary."

It felt like having a built-in tour guide for the site.

Step 3: Integrating Subscriptions

This was the real game-changer. I connected the chatbot to my subscription system so it could recognize which plan a user was on.

- Free plan users saw prompts like:
  "You're currently on a free plan. Want to unlock the full video library? Click here to upgrade."
- Premium users received tailored support:
  "Thanks for being a Premium member! Would you like me to show you the new videos added this week?"

In other words, the chatbot wasn't just supporting users—it was **personalizing their journey and encouraging upgrades.**

Step 4: Real-World Testing

I invited friends, family, and educators to try it out. Their feedback was eye-opening:

- "It feels like a real tutor guiding me through the site."
- "I don't have to click around—I just ask and get the answer."
- "This actually makes me want to stay longer."

Their reactions proved that the bot wasn't just functional—it was engaging and helpful.

Step 5: Transformation Into an AI Buddy

With these connections in place, the chatbot stopped being a widget in the corner and became the **heartbeat of the site**. It now:

- Guides first-time visitors
- Answers student questions in real-time
- Suggests resources to parents and teachers
- Acts as both a tutor and a sales assistant

It turned *YourAIStudyBuddy.com* from a static ecommerce site into an **interactive AI-powered learning platform.**

Final Thoughts

Building the chatbot in Chapter 20 was an achievement. But **connecting it in Chapter 21 was the breakthrough**. That's when it stopped being a tool and started being a companion—an AI buddy who helps learners explore, understand, and grow.

When I see students chatting with it today, I know this isn't just a feature. It's the heart of my vision: a place where AI doesn't just sit in the background, but actively teaches, supports, and inspires.

Adding Multi-Agent AI Platform: Complete Implementation Guide

Building Specialized AI Agents for Educational Excellence

Chapter 22: Understanding Multi-Agent Architecture

The evolution from single-chatbot systems to sophisticated multi-agent platforms represents a fundamental shift in how we approach artificial intelligence implementation for educational purposes. When you initially considered adding a customer support bot to your website, you were thinking in terms of a single-purpose solution designed to handle basic inquiries and provide automated assistance. However, the landscape of AI-powered education demands a more nuanced approach that recognizes the diverse learning needs of students, educators, parents, and business professionals who will interact with your platform.

Multi-agent architecture operates on the principle that different specialized agents can provide more targeted, effective assistance than a single generalist bot attempting to handle all scenarios. Think of this approach like a well-organized school where you have different teachers for mathematics, writing, and specialized subjects. Each educator brings deep expertise in their domain, understands the common challenges students face in that area, and can provide instruction tailored to the specific requirements of their subject matter. Your AI agents function similarly, with each one trained and optimized for particular types of interactions and learning objectives.

The mathematical complexity of calculus problems requires different processing approaches than creative writing assignments, which in turn demand different capabilities than marketing strategy development. A math agent needs to understand numerical relationships, formula applications, step-by-step problem-solving methodologies, and the ability to break down complex equations into manageable components. Meanwhile, a writing agent must grasp grammar rules, style conventions, narrative structure, argumentation techniques, and the subtle nuances of tone and voice that make communication effective.

When designing your multi-agent system for youraistudybuddy.com, you are essentially creating a virtual faculty of specialized instructors, each with their own expertise, teaching methodologies, and approaches to student engagement. This specialization allows for more accurate responses, better user experiences, and ultimately more effective learning outcomes. Students struggling with algebraic concepts can interact with an agent specifically designed to understand mathematical thinking patterns, while those working on essay assignments can engage with an agent trained in composition techniques and writing best practices.

The architectural foundation of your multi-agent system also enables scalability and future expansion. Rather than rebuilding a monolithic system when you want to add new capabilities, you can develop additional specialized agents that integrate seamlessly with your existing platform. This modular approach means you could eventually add agents for science, history, foreign languages, or any other subject area without disrupting the functionality of your current agents.

Understanding user routing becomes crucial in multi-agent systems. Your platform needs intelligent mechanisms to direct users to the most appropriate agent based on their stated needs, the type of questions they ask, or the subject matter they are working with. This routing system acts as a digital receptionist, ensuring that each user connects with the agent best equipped to help them achieve their learning goals.

The conversational context management in multi-agent systems also differs significantly from single-bot implementations. Each agent needs to maintain awareness of its specialized domain while having the ability to recognize when a user's needs might be better served by a different agent. This requires sophisticated handoff mechanisms that preserve conversation history and context while transitioning users between agents smoothly and naturally.

Chapter 23: Planning Your AI Agent Ecosystem

Strategic planning forms the cornerstone of successful multi-agent implementation, and your approach to designing the AI ecosystem for youraistudybuddy.com must consider both current user needs and future expansion possibilities. The process begins with comprehensive analysis of your target audience segments and their specific learning requirements. Students approaching your platform will arrive with varying levels of mathematical understanding, from basic arithmetic concepts to advanced calculus and statistics. Similarly, writing assistance needs span from elementary grammar correction to sophisticated research paper development and professional communication skills.

Your educational background research through the creation of your comprehensive AI dictionary provides invaluable insights into the knowledge gaps and learning challenges that different user groups face. This research phase should inform the specialization boundaries for each agent, ensuring that the division of responsibilities creates clear value propositions for users while avoiding unnecessary overlap or confusion about which agent to consult for specific types of assistance.

The mathematical agent specialization requires careful consideration of curriculum standards and common learning progressions. Elementary mathematics focuses on number sense, basic operations, and foundational problem-solving skills. Middle school mathematics introduces algebraic thinking, geometric concepts, and data analysis fundamentals. High school mathematics encompasses advanced algebra, trigonometry, pre-calculus concepts, and introduction to statistical thinking. Your basic math agent should be designed to handle these foundational through intermediate concepts with patient, step-by-step explanations that build understanding progressively.

Advanced mathematics presents entirely different challenges and requires specialized knowledge in calculus, advanced statistics, linear algebra, differential equations, and mathematical proof techniques. Students at this level need agents capable of discussing complex theoretical concepts, providing guidance on proof construction, explaining the relationships between different mathematical domains, and helping with research-level problem-solving approaches. The advanced math agent must be equipped with extensive knowledge of mathematical literature, theorem applications, and the ability to guide students through sophisticated analytical thinking processes.

Writing instruction encompasses an equally broad spectrum of skills and applications. Basic writing assistance includes grammar correction, sentence structure improvement, paragraph organization, and fundamental composition skills. Students need help understanding how to develop ideas clearly, organize thoughts logically, and express themselves effectively in written form. The writing agent must be capable of providing constructive feedback on draft materials, suggesting improvements that enhance clarity and impact, and teaching writing techniques that students can apply across different contexts.

Advanced writing instruction moves into specialized domains including academic research paper development, professional communication, creative writing techniques, and subject-specific writing requirements. Graduate students working on thesis projects need different types of assistance than business professionals developing marketing materials or creative writers crafting narratives. Your writing agent ecosystem should be designed to recognize these different contexts and provide appropriate guidance for each situation.

The marketing agent represents a unique addition to educational AI platforms, reflecting your understanding that business education and entrepreneurial skill development are increasingly important components of comprehensive education. This agent needs knowledge of marketing principles, strategic thinking frameworks, digital marketing techniques, content creation strategies, and business communication best practices. The marketing agent should be capable of helping users develop business plans, create marketing campaigns, understand target audience analysis, and develop professional presentation skills.

Integration planning requires careful attention to how these specialized agents will work together within your platform ecosystem. Users may need assistance that spans multiple domains, such as a student working on a research paper that requires statistical analysis of collected data. The mathematical and writing agents need coordination mechanisms that allow them to collaborate effectively on such cross-disciplinary projects while maintaining their specialized expertise and approaches.

User journey mapping becomes essential for understanding how different user types will interact with your multi-agent system. Traditional students may primarily use mathematical and writing agents for homework assistance and skill development. Adult learners returning to education might need more comprehensive support across all agent types as they develop new professional capabilities. Business owners exploring your platform may focus primarily on marketing agent interactions while occasionally needing writing assistance for professional communications.

Chapter 24: Setting Up Advanced Botpress Configurations

The transition from single-agent to multi-agent Botpress configuration requires fundamental changes to your platform architecture and administrative approach. Your existing customer support bot provides a foundation of Botpress familiarity, but supporting multiple specialized agents demands more sophisticated workspace organization, conversation flow management, and integration strategies that scale effectively as your platform grows.

Workspace architecture in Botpress becomes significantly more important when managing multiple agents simultaneously. Rather than treating each agent as a completely separate bot, advanced configuration involves creating a unified workspace that allows for centralized management while maintaining distinct agent personalities and capabilities. This approach provides administrative efficiency while ensuring consistent user experiences across all agent interactions.

The initial workspace setup process involves establishing a hierarchical structure that reflects your agent specializations and anticipated usage patterns. Your primary workspace should contain global configurations that apply across all agents, including user authentication systems, general conversation policies, escalation procedures, and integration settings with your WordPress platform. Secondary configurations then handle agent-specific requirements such as specialized knowledge bases, conversation templates, and response pattern libraries.

Knowledge base architecture represents one of the most critical configuration decisions in your multi-agent system. Each specialized agent requires access to domain-specific information while maintaining the ability to reference general educational resources and your comprehensive AI dictionary content. The mathematical agents need extensive formula libraries, problem-solving methodologies, worked example databases, and connections to computational resources that can assist with complex calculations and graphical representations.

Writing agents require different knowledge base components including style guides, grammar references, writing technique libraries, example compositions across different genres and academic levels, and citation format databases. The knowledge base must also include information about different writing contexts, audience considerations, and purpose-driven communication strategies that help users develop appropriate voice and tone for their specific writing objectives.

Marketing agent knowledge bases encompass business strategy frameworks, marketing methodology references, case study libraries, industry trend information, and tools for market analysis and campaign development. This agent needs access to current information about digital marketing techniques, social media strategies, content creation best practices, and business communication standards that reflect contemporary professional expectations.

Conversation flow design becomes substantially more complex in multi-agent environments because each agent needs distinct interaction patterns while maintaining consistency with overall platform experience. Mathematical agents should be configured to encourage step-by-step problem-solving approaches, ask clarifying questions about problem context and requirements, and provide detailed explanations that build understanding progressively. The conversation flows should include checkpoints where the agent confirms user understanding before proceeding to more complex concepts.

Writing agent conversation flows should be designed to understand project scope, audience requirements, writing purpose, and current skill level before providing assistance. These flows need branching logic that adapts to different types of writing projects, from creative assignments to academic research papers to professional communications. The agent should be configured to ask for draft materials when appropriate, provide structured feedback, and offer revision suggestions that align with the specific writing goals.

Marketing agent flows require understanding of business context, target market characteristics, budget considerations, and strategic objectives before providing marketing guidance. The conversation design should include discovery processes that help the agent understand the user's business situation, competitive environment, and marketing experience level to provide appropriately tailored advice and recommendations.

Integration protocols between agents become essential when users need assistance that spans multiple domains. Your Botpress configuration must include handoff mechanisms that preserve conversation context while transitioning users between agents smoothly. These protocols should maintain user preferences, conversation history, and project-specific information that enables continued assistance without requiring users to repeat background information.

Natural Language Processing customization for each agent requires attention to domain-specific terminology, common question patterns, and specialized vocabulary that users employ when seeking assistance in different subject areas. Mathematical agents need understanding of mathematical notation, problem-solving language, and numerical concept terminology. Writing agents require sensitivity to style terminology, grammatical concepts, and composition vocabulary. Marketing agents need familiarity with business terminology, strategic language, and industry-specific communications.

Chapter 25: Creating the AI Math Agent

The development of your AI Math Agent represents the foundation of specialized educational assistance on youraistudybuddy.com, serving students from elementary arithmetic through advanced high school mathematics. This agent must be designed to understand not just mathematical procedures, but the underlying conceptual frameworks that make mathematics comprehensible and applicable to real-world situations. The creation process involves careful attention to pedagogical approaches that build understanding progressively while accommodating different learning styles and mathematical backgrounds.

Mathematical concept organization within the agent requires systematic structuring that reflects how mathematical understanding develops naturally. Elementary concepts including number sense, basic operations, fractions, and decimal representations form the foundation that supports more advanced topics like algebraic thinking, geometric relationships, and statistical reasoning. Your agent configuration must recognize these prerequisite relationships and be capable of identifying when users need foundational review before tackling more complex problems.

The problem-solving methodology embedded in your math agent should reflect best practices in mathematics education, emphasizing understanding over memorization and encouraging students to develop analytical thinking skills that transfer across different mathematical domains. When students present word problems, the agent should guide them through systematic approaches including problem interpretation, relevant information identification, strategy selection, solution implementation, and answer verification. This structured approach helps students develop confidence in their mathematical reasoning abilities.

Step-by-step explanation capabilities represent one of the most valuable features of your math agent, providing detailed breakdowns of solution processes that help students understand not just what to do, but why particular approaches work effectively. These explanations should be tailored to the student's indicated grade level and mathematical background, using appropriate vocabulary and examples that connect to familiar concepts and experiences.

Visual representation support enhances the math agent's effectiveness by helping students understand geometric concepts, graphical relationships, and data visualization techniques. While the agent may not generate complex graphics directly, it should be capable of describing visual elements clearly, providing instructions for creating graphs and diagrams, and explaining how visual representations connect to numerical and algebraic concepts.

Error analysis and correction guidance helps students learn from mistakes and develop stronger mathematical reasoning skills. When students present incorrect solutions or express confusion about particular concepts, the agent should be configured to identify common error patterns, provide gentle correction with clear explanations, and suggest practice strategies that address underlying misconceptions or knowledge gaps.

Homework assistance protocols ensure that the math agent provides educational support rather than simply completing assignments for students. The configuration should emphasize guided discovery, asking questions that help students think through problems independently, providing hints and suggestions rather than complete solutions, and encouraging students to explain their thinking processes to reinforce learning.

Mathematical vocabulary development receives special attention in the agent design because mathematical communication requires precision and clarity that differs from everyday language usage. The agent should be equipped to explain mathematical terminology clearly, provide examples of proper usage, and help students develop comfort with mathematical notation and symbolic representation systems.

Real-world application connections help students understand the relevance and utility of mathematical concepts beyond classroom contexts. The math agent should include examples and scenarios that demonstrate how mathematical thinking applies to everyday situations, future academic work, and career applications that might interest students with different goals and interests.

Differentiated instruction capabilities allow the agent to adapt its teaching approaches to accommodate different learning preferences and mathematical strengths. Some students benefit from concrete examples and manipulative-based explanations, while others prefer abstract reasoning and algebraic approaches. The agent should be capable of recognizing when alternative explanation methods might be more effective and adjusting its communication style accordingly.

Assessment and progress monitoring features help students understand their mathematical development and identify areas that need additional attention. The agent should be configured to track the types of problems students work on, recognize patterns in questions and difficulties, and provide feedback about learning progress that encourages continued mathematical exploration and skill development.

Chapter 26: Developing the Advanced Math Agent

The Advanced Math Agent serves students, educators, and professionals working with sophisticated mathematical concepts that extend beyond traditional high school curriculum into college-level and specialized mathematical domains. This agent requires substantially different knowledge bases, reasoning capabilities, and instructional approaches compared to the basic math agent, reflecting the increased abstraction, theoretical complexity, and application sophistication that characterizes advanced mathematical work.

Calculus instruction represents a primary focus area for the advanced math agent, encompassing differential calculus concepts including limits, derivatives, applications of differentiation, integral calculus covering integration techniques and applications, and multivariable calculus extending these concepts to higher dimensions. The agent must be equipped to explain theoretical foundations while providing practical problem-solving guidance that helps students understand both computational techniques and conceptual significance.

Linear algebra concepts require specialized knowledge of vector spaces, matrix operations, determinants, eigenvalues, eigenvectors, and their applications in various mathematical and applied contexts. The advanced math agent needs capability to explain abstract algebraic structures while connecting these concepts to concrete applications in computer science, engineering, physics, and other fields where linear algebra provides essential analytical tools.

Statistical analysis and probability theory demand comprehensive understanding of descriptive statistics, probability distributions, hypothesis testing, regression analysis, and advanced statistical methods used in research and data analysis contexts. The agent should be capable of guiding users through statistical reasoning processes, helping them select appropriate analytical techniques, and interpreting statistical results in meaningful ways.

Differential equations instruction encompasses ordinary differential equations, partial differential equations, and their applications in modeling dynamic systems across scientific and engineering disciplines. The advanced math agent must understand solution techniques while explaining the conceptual significance of differential equations in describing real-world phenomena and predictive modeling applications.

Mathematical proof techniques represent a crucial component of advanced mathematics that requires different instructional approaches than computational problem-solving. The agent needs expertise in direct proof methods, proof by contradiction, mathematical induction, and other logical reasoning structures that form the foundation of mathematical argumentation. Students working on proof-based assignments need guidance in logical structure, rigorous reasoning, and clear mathematical communication.

Research-level problem solving involves helping users approach mathematical problems that may not have standard solution procedures or well-defined answer formats. The advanced math agent should be configured to help users break down complex problems, identify relevant theoretical frameworks, develop solution strategies, and evaluate the validity and significance of their approaches and results.

Mathematical modeling capabilities enable the agent to assist users in translating real-world situations into mathematical representations that can be analyzed using appropriate mathematical tools and techniques. This involves understanding how to identify key variables, establish relationships between quantities, select suitable mathematical frameworks, and interpret mathematical results in the context of original problem situations.

Advanced computational support helps users leverage technology tools appropriately while maintaining focus on mathematical understanding and reasoning. The agent should be familiar with computer algebra systems, statistical software, numerical analysis techniques, and other technological resources that support advanced mathematical work without replacing mathematical thinking and analysis.

Interdisciplinary applications showcase how advanced mathematics connects to other academic disciplines and professional fields. The agent should understand how mathematical concepts apply in physics, engineering, economics, biology, computer science, and other areas where mathematical modeling and analysis provide essential insights and problem-solving capabilities.

Graduate-level assistance extends the agent's capabilities to support students working on master's and doctoral-level mathematical coursework, comprehensive examinations, and research projects. This requires understanding of specialized mathematical literature, current research directions, and the ability to provide guidance on advanced topics that may be at the frontier of mathematical knowledge.

Professional development support helps educators, researchers, and professionals who need to expand their mathematical knowledge for career advancement or project requirements. The agent should be capable of providing efficient pathways for learning new mathematical concepts, connecting new knowledge to existing expertise, and supporting practical applications in professional contexts.

Chapter 27: Building the AI Writing Agent

The AI Writing Agent development process focuses on creating comprehensive support for written communication across educational levels and professional contexts, recognizing that effective writing skills are essential for academic success and career advancement. This agent must be designed to understand the complexity of written communication, from basic grammar and sentence structure through sophisticated argumentation and professional document creation.

Grammar and mechanics instruction forms the foundation of the writing agent's capabilities, providing support for sentence construction, punctuation usage, verb tense consistency, subject-verb agreement, and other fundamental elements of clear written communication. The agent should be equipped to identify common grammatical errors, explain correction strategies, and help users develop proofreading skills that improve the technical accuracy of their writing.

Composition structure guidance helps users organize their ideas effectively, whether they are working on paragraph development, essay organization, research paper structure, or professional document formatting. The agent needs knowledge of various organizational patterns including chronological, compare-and-contrast, cause-and-effect, problem-solution, and argumentative structures that serve different communication purposes.

Writing process support acknowledges that effective writing involves multiple stages including brainstorming, drafting, revising, and editing. The agent should be configured to provide assistance at each stage, helping users generate ideas, organize thoughts, develop initial drafts, identify areas for improvement, and polish final documents. This process-oriented approach helps users develop sustainable writing practices that improve over time.

Audience awareness instruction helps users understand how to adapt their writing style, vocabulary, tone, and content organization to meet the needs and expectations of different readers. Academic writing requires different approaches than professional communications, which differ from creative writing or personal expression. The agent should guide users in making appropriate choices for their intended audience and communication purpose.

Research and citation support provides essential assistance for academic and professional writing that incorporates external sources. The agent needs knowledge of various citation formats including MLA, APA, Chicago, and others, helping users integrate source material effectively while maintaining proper attribution and avoiding plagiarism issues. This includes guidance on paraphrasing, quotation integration, and bibliography preparation.

Argumentative writing instruction helps users develop logical reasoning skills and persuasive communication techniques. The agent should understand how to guide users in developing clear thesis statements, supporting arguments with appropriate evidence, addressing counterarguments effectively, and creating compelling conclusions that reinforce their main points.

Creative writing support encompasses fiction, poetry, personal narrative, and other forms of creative expression that require different skills than academic or professional writing. The agent should be equipped to provide guidance on character development, plot structure, narrative techniques, descriptive writing, and other creative elements while encouraging users to develop their unique voice and style.

Professional communication assistance includes business letters, reports, proposals, presentations, email communications, and other workplace writing requirements. The agent needs understanding of professional conventions, appropriate tone and formality levels, document formatting standards, and communication strategies that achieve business objectives effectively.

Revision and editing guidance helps users develop critical evaluation skills for their own writing, teaching them to identify areas for improvement, reorganize content for better clarity and impact, and refine their expression to communicate more effectively. This includes instruction on global revision strategies that address content and organization as well as local editing techniques that improve sentence-level clarity.

Writing across disciplines acknowledges that different academic and professional fields have distinct writing conventions, vocabulary, and communication expectations. The agent should be capable of providing guidance appropriate for scientific writing, humanities scholarship, business communications, technical documentation, and other specialized contexts that require field-specific knowledge and approaches.

Collaborative writing support helps users participate effectively in group writing projects, providing guidance on coordination strategies, revision processes, and integration techniques that produce cohesive documents when multiple writers contribute to the final product.

Chapter 28: Constructing the AI Marketing Agent

The AI Marketing Agent represents a unique addition to educational AI platforms, reflecting the growing importance of marketing literacy and entrepreneurial skills across various educational and professional contexts. This agent must be designed to serve students learning business concepts, professionals developing marketing campaigns, small business owners seeking strategic guidance, and educators teaching marketing principles and practices.

Marketing fundamentals instruction begins with core concepts including market research, target audience identification, value proposition development, competitive analysis, and strategic positioning that form the foundation of effective marketing planning. The agent should be equipped to explain these concepts clearly while providing practical guidance for applying marketing principles in real-world situations.

Digital marketing expertise encompasses contemporary marketing channels and techniques including social media marketing, content marketing, search engine optimization, email marketing, pay-per-click advertising, and online analytics that dominate current marketing practice. The agent needs current knowledge of platform-specific strategies, best practices, and measurement techniques that help users achieve their marketing objectives effectively.

Campaign development support helps users plan, implement, and evaluate marketing campaigns across various channels and objectives. This includes guidance on campaign goal setting, budget allocation, timeline development, creative content planning, implementation coordination, and performance measurement that enables continuous improvement and strategic refinement.

Brand development assistance provides guidance on creating and maintaining consistent brand identity across marketing communications and customer touchpoints. The agent should understand logo design principles, color psychology, messaging consistency, brand voice development, and reputation management strategies that build strong brand recognition and customer loyalty.

Content creation guidance helps users develop marketing materials including blog posts, social media content, advertising copy, website content, email newsletters, and other communications that engage target audiences effectively. The agent needs knowledge of content marketing principles, audience engagement strategies, and platform-specific best practices that maximize content impact and reach.

Market research methodologies provide users with tools and techniques for gathering and analyzing market information that informs strategic decision-making. The agent should be capable of explaining primary and secondary research methods, survey design principles, data collection techniques, and analysis approaches that provide actionable insights for marketing planning.

Strategic planning support helps users develop comprehensive marketing strategies that align with business objectives and available resources. This includes guidance on situation analysis, objective setting, strategy selection, tactical planning, budget development, and implementation timelines that create realistic and achievable marketing plans.

Performance measurement and analytics instruction helps users understand how to track marketing effectiveness, interpret performance data, and make data-driven decisions that improve marketing results. The agent needs knowledge of key performance indicators, analytics tools, measurement frameworks, and optimization strategies that enable continuous improvement.

Customer relationship management guidance provides instruction on building and maintaining positive customer relationships through effective communication, service excellence, feedback management, and loyalty program development that creates long-term customer value and business sustainability.

Entrepreneurial marketing addresses the unique challenges and opportunities that small businesses and startups face when competing with limited resources against established competitors. The agent should understand bootstrapping techniques, guerrilla marketing strategies, partnership development, and creative approaches that maximize marketing impact while minimizing costs.

Ethical marketing principles ensure that users understand responsible marketing practices that respect consumer rights, avoid deceptive practices, and contribute positively to community well-being. The agent should be equipped to discuss issues including truthful advertising, privacy protection, target audience appropriateness, and sustainable marketing practices.

Industry-specific applications recognize that marketing strategies and tactics vary significantly across different industries and market contexts. The agent should understand how marketing principles apply differently in retail, services, manufacturing, non-profit, technology, healthcare, and other sectors that have distinct customer bases and marketing requirements.

Chapter 29: Agent Integration and WordPress Implementation

The integration of multiple specialized AI agents into your WordPress-based youraistudybuddy.com platform requires sophisticated technical implementation and user experience design that seamlessly connects agent capabilities with website functionality. This integration process involves embedding agent interfaces, managing user routing between agents, maintaining conversation continuity, and ensuring optimal performance across all platform components.

WordPress integration architecture begins with establishing the technical framework that connects Botpress agents with your website's content management system and user interface. This involves configuring embed codes, establishing secure communication protocols, managing user authentication between platforms, and creating database connections that support user progress tracking and conversation history storage.

Agent selection interface design creates the user-facing mechanisms that allow visitors to choose appropriate agents for their specific needs and learning objectives. This interface should provide clear descriptions of agent capabilities, suggest appropriate agents based on user-indicated needs, and enable smooth transitions between agents when users need assistance across multiple domains.

User routing logic determines how the platform directs users to appropriate agents based on their stated needs, question types, subject matter, and user profile information. This system should be intelligent enough to recognize when users might benefit from different agents than they initially selected while providing override options that respect user preferences and choices.

Conversation handoff mechanisms enable users to move between agents while preserving conversation context, user preferences, and project-specific information that enables continued assistance without repetitive background explanations. These handoffs should feel natural and seamless from the user perspective while maintaining technical reliability and data security.

Authentication integration ensures that user accounts, subscription levels, progress tracking, and usage analytics work consistently across all agent interactions. This involves connecting Botpress user management with WordPress user systems, maintaining session continuity between agents, and ensuring that subscription-based access controls function properly across all platform components.

Performance optimization addresses the technical challenges of supporting multiple concurrent agent conversations while maintaining responsive website performance and reliable user experiences. This includes load balancing, server resource management, caching strategies, and monitoring systems that ensure consistent platform availability and response times.

Mobile responsiveness ensures that agent interfaces function effectively across desktop, tablet, and smartphone devices that students and professionals use to access educational resources. The implementation must provide consistent functionality and user experience regardless of device type while accommodating different screen sizes and interaction methods.

Database integration supports user progress tracking, conversation logging, usage analytics, and subscription management across all agent interactions. This requires establishing data structures that efficiently store and retrieve multi-agent conversation histories, user preferences, learning progress indicators, and administrative information needed for platform management.

Security protocols protect user privacy, conversation confidentiality, and platform integrity while enabling the collaborative functionality that makes multi-agent systems valuable for educational applications. This includes data encryption, secure transmission protocols, user access controls, and compliance with educational privacy regulations.

Analytics integration provides administrators with comprehensive insights into platform usage patterns, agent effectiveness, user satisfaction levels, and business metrics that inform platform development and marketing strategies. This system should track metrics across all agents while providing agent-specific performance indicators that guide improvement efforts.

Customization options allow administrators to modify agent personalities, update knowledge bases, adjust conversation flows, and implement new features without requiring complete system rebuilds. This flexibility enables platform evolution and responsiveness to user feedback and changing educational needs.

Backup and recovery systems ensure platform reliability and data protection through automated backup procedures, disaster recovery protocols, and system monitoring that prevents data loss and minimizes service disruptions that could impact user learning experiences.

Chapter 30: User Experience and Interface Design

User experience design for multi-agent educational platforms requires careful attention to navigation clarity, interface consistency, and interaction patterns that support effective learning while minimizing confusion and technical barriers. Your youraistudybuddy.com platform must provide intuitive access to specialized agent capabilities while maintaining coherent overall user experience that accommodates diverse user types and technical skill levels.

Navigation architecture establishes clear pathways for users to access different agents, understand platform capabilities, and move efficiently between different types of assistance without losing context or becoming disoriented. The navigation system should provide consistent access to all agents while highlighting the most relevant options based on user profiles and current activities.

Visual design consistency ensures that all agent interfaces share common design elements, color schemes, typography, and interaction patterns while allowing for specialization that reflects each agent's unique purpose and personality. This balance between consistency and specialization helps users develop familiarity with platform conventions while recognizing distinct agent capabilities.

Onboarding processes introduce new users to platform capabilities, help them identify appropriate agents for their needs, and provide initial guidance that enables successful first interactions. The onboarding should be efficient enough to avoid frustrating eager learners while comprehensive enough to prevent confusion and ineffective platform usage.

Agent personality development creates distinct but complementary interaction styles that reflect each agent's specialized knowledge and teaching approaches. The math agents might adopt systematic, step-by-step communication patterns, while writing agents could emphasize collaborative, process-oriented interactions, and marketing agents might focus on strategic, goal-oriented discussions.

Conversation interface design addresses the technical and aesthetic elements of agent interactions including message formatting, input methods, file sharing capabilities, and visual elements that enhance communication effectiveness. These interfaces should support the specific needs of each agent type while maintaining overall platform coherence.

Progress visualization helps users understand their learning development across different subject areas and agent interactions. This might include skill progression indicators, concept mastery tracking, goal achievement displays, and other visual elements that motivate continued engagement and provide clear feedback about educational progress.

Accessibility considerations ensure that the platform serves users with diverse abilities and technical access requirements. This includes keyboard navigation support, screen reader compatibility, adjustable font sizes, color contrast options, and other accommodations that make educational AI accessible to all potential users.

Error handling and recovery systems help users navigate technical difficulties, connectivity issues, and misunderstandings that might occur during agent interactions. These systems should provide clear error messages, suggest resolution strategies, and enable users to continue their learning activities with minimal disruption.

Customization options allow users to adjust interface elements, notification preferences, agent interaction styles, and other platform features to match their individual learning preferences and technical requirements. This personalization capability enhances user satisfaction and platform effectiveness.

Cross-device synchronization ensures that users can begin conversations on one device and continue them on another while maintaining conversation history, progress tracking, and preference settings. This flexibility accommodates modern learning patterns where users switch between computers, tablets, and smartphones throughout their educational activities.

Search and discovery features help users find previous conversations, locate specific information within their learning history, and access relevant resources across different agent interactions. These features become increasingly important as users accumulate extensive conversation histories with multiple agents.

Collaboration tools support group learning activities, parent-child interactions, teacher-student communications, and other multi-user scenarios that enhance the educational value of agent interactions. These tools should maintain privacy and security while enabling appropriate sharing and communication.

Chapter 31: Testing and Quality Assurance

Comprehensive testing protocols ensure that your multi-agent platform functions reliably across diverse user scenarios, technical environments, and educational contexts before launching to your target audience of students, educators, parents, and professionals. The testing process must address individual agent functionality, cross-agent integration, WordPress platform compatibility, and user experience quality across different devices and usage patterns.

Individual agent testing begins with validating each agent's knowledge base accuracy, response appropriateness, and ability to handle typical user queries within their specialized domains. Mathematical agents require testing with problems across different difficulty levels, various mathematical topics, and different problem presentation formats to ensure comprehensive coverage and accurate assistance.

Conversation flow testing examines how effectively each agent guides users through typical interaction patterns including problem-solving sequences, explanation requests, clarification needs, and error correction scenarios. This testing should include both successful interaction paths and challenging situations where users provide incomplete information, ask unclear questions, or need additional support.

Knowledge base validation ensures that agent responses reflect current best practices in their respective domains, align with educational standards and curriculum requirements, and provide accurate information that supports effective learning. This is particularly important for mathematical formulas, writing conventions, and marketing strategies that evolve over time.

Cross-agent integration testing verifies that handoff mechanisms work smoothly when users need assistance that spans multiple agent specializations. This includes testing conversation context preservation, user preference maintenance, and the overall user experience during transitions between agents with different personalities and approaches.

WordPress integration testing validates that agent embed codes function properly within your website's theme and plugin environment, user authentication systems work consistently across platform components, and page loading performance remains acceptable when agent interfaces are active.

Mobile compatibility testing ensures that agent interfaces function effectively on smartphones and tablets with different screen sizes, operating systems, and browser configurations. This testing should address touch interface usability, conversation readability, input method functionality, and overall user experience quality across mobile devices.

Performance testing evaluates platform response times, concurrent user capacity, server resource utilization, and system reliability under various load conditions. This testing helps identify potential bottlenecks and ensures that platform performance remains acceptable as user numbers grow.

Security testing validates user data protection, conversation privacy, authentication security, and protection against common web application vulnerabilities. This is particularly important for educational platforms that may serve minors and must comply with privacy regulations like COPPA and FERPA.

Accessibility testing ensures that platform features work effectively with screen readers, keyboard navigation, and other assistive technologies used by individuals with disabilities. This testing should address all agent interfaces and platform features to ensure comprehensive accessibility compliance.

User acceptance testing involves real users from your target audience interacting with agents under realistic conditions to identify usability issues, confusing interface elements, and gaps between agent capabilities and user expectations. This testing provides crucial feedback for refining agent personalities and platform functionality.

Content accuracy review examines agent responses for factual correctness, pedagogical effectiveness, and alignment with educational best practices. This review should involve subject matter experts who can evaluate whether agent assistance actually supports effective learning and skill development.

Error handling testing validates how effectively the platform manages technical failures, network connectivity issues, user input errors, and other exceptional conditions that might occur during normal platform usage. This testing ensures that users receive helpful error messages and can recover from problems without losing their work or becoming frustrated with the platform.

Chapter 32: Performance Optimization and Scaling

Platform performance optimization becomes critically important as your youraistudybuddy.com multi-agent system grows to serve thousands of concurrent users across different time zones and usage patterns. Educational platforms face unique performance challenges because users expect immediate responses during homework sessions, study periods, and assignment deadlines when platform traffic typically peaks.

Server architecture planning addresses the technical infrastructure requirements for supporting multiple Botpress agents, WordPress content management, user authentication, conversation logging, and analytics processing simultaneously. This infrastructure must be designed to scale efficiently as user numbers grow without requiring complete system rebuilds or extended service disruptions.

Load balancing strategies distribute user requests across multiple servers to prevent any single system component from becoming overwhelmed during peak usage periods. Educational platforms often experience traffic spikes during evening hours when students work on homework, weekend study sessions, and periods just before assignment deadlines when usage can increase dramatically.

Database optimization ensures that user conversation histories, progress tracking data, subscription information, and analytics data can be stored and retrieved efficiently even as these databases grow to contain millions of interactions across thousands of users. This includes indexing strategies, query optimization, and archival procedures that maintain performance over time.

Content delivery network implementation accelerates platform loading times by distributing static resources geographically closer to users regardless of their location. This is particularly important for international users who might otherwise experience slow response times that interfere with effective learning interactions.

Caching strategies reduce server processing requirements by storing frequently requested information in readily accessible formats. This includes agent response caching for common questions, user interface element caching, and database query result caching that reduces the computational overhead of serving repetitive requests.

Memory management optimization addresses how efficiently the platform utilizes server memory resources while maintaining multiple agent instances, user sessions, and background processes simultaneously. Effective memory management prevents system slowdowns and crashes that could disrupt user learning experiences during critical study periods.

API rate limiting protects your platform from overuse while ensuring fair access for all users during high-traffic periods. This involves implementing intelligent throttling mechanisms that prioritize educational activities while preventing any single user or automated system from overwhelming platform resources.

Monitoring and alerting systems provide real-time visibility into platform performance metrics, user experience quality, and potential issues before they impact user satisfaction. These systems should track response times, error rates, user satisfaction indicators, and resource utilization patterns that inform optimization decisions.

Scalability planning establishes procedures for expanding platform capacity as user numbers grow from hundreds to thousands to potentially tens of thousands of active learners. This planning should address both technical scaling requirements and operational procedures that maintain service quality during growth periods.

Resource allocation optimization ensures that server computing power, memory, and bandwidth are distributed effectively across different platform components based on actual usage patterns and performance requirements. Educational platforms may need to prioritize mathematical computation resources during homework periods while emphasizing writing support during research paper seasons.

International performance considerations address the unique challenges of serving users across different countries, time zones, and internet infrastructure quality levels. This includes geographic server distribution, language processing optimization, and connectivity accommodation that provides consistent experiences regardless of user location.

Mobile performance optimization recognizes that many students access educational resources primarily through smartphones and tablets that have different processing capabilities and network connectivity constraints compared to desktop computers. Mobile optimization includes interface streamlining, data usage minimization, and offline capability development where appropriate.

Cost optimization strategies balance platform performance requirements with operational budget constraints, particularly important during the early growth phases when revenue may not yet cover premium infrastructure costs. This involves intelligent resource provisioning, usage-based scaling, and cost monitoring that maintains financial sustainability.

Platform reliability engineering establishes redundancy, backup systems, and recovery procedures that maintain educational continuity even when individual system components experience failures or require maintenance. Students cannot afford to lose access to homework assistance during critical study periods, making reliability a top priority.

Update and maintenance procedures ensure that agent knowledge bases remain current, security patches are applied promptly, and new features can be deployed without disrupting ongoing user activities. Educational platforms require particularly careful change management because disruptions during exam periods or assignment deadlines can significantly impact user success.

Chapter 33: Launch Strategy and Ongoing Management

The strategic launch of your multi-agent youraistudybuddy.com platform requires careful coordination of technical deployment, marketing communications, user onboarding, and support systems that collectively create positive first impressions while establishing sustainable growth patterns. Your launch strategy must accommodate the diverse needs of students, educators, parents, and professionals while building the user base necessary for long-term platform success.

Pre-launch preparation involves finalizing all technical components, completing comprehensive testing protocols, preparing user documentation and help resources, training any support personnel, and establishing monitoring systems that will track platform performance and user satisfaction from day one. This preparation phase should include soft launch activities with limited user groups that provide final validation before full public availability.

Target audience segmentation guides your launch communications and onboarding processes by recognizing that different user types have distinct needs, communication preferences, and adoption patterns. Students may discover your platform through social media and peer recommendations, while educators might require more detailed information about curriculum alignment and classroom integration possibilities.

Marketing communication strategy leverages your existing authority through your published AI dictionary and technical books to establish credibility and attract early adopters who recognize the value of comprehensive AI education resources. Your international dictionary sales provide valuable social proof that can accelerate platform adoption among educators and institutions.

Onboarding sequence design creates positive first experiences that help new users understand platform capabilities, select appropriate agents for their needs, and achieve early success that encourages continued engagement. Different user types may benefit from customized onboarding paths that highlight the most relevant features and capabilities for their specific goals.

Pricing strategy implementation involves activating your planned subscription tiers, token limitation systems, and payment processing integration while ensuring that free trial periods provide sufficient value demonstration to encourage conversion to paid subscriptions. The pricing strategy should reflect the significant value that specialized AI tutoring provides compared to traditional educational support options.

Customer support systems must be ready to handle technical questions, usage guidance, billing inquiries, and educational support requests from users who may have limited technical experience but high expectations for educational assistance quality. Support documentation should address common questions while personal support channels handle more complex issues.

Content marketing initiatives showcase platform capabilities through blog posts, social media content, webinars, and other educational resources that demonstrate agent effectiveness while providing genuine value to potential users. This content can highlight success stories, provide educational tips, and establish thought leadership in AI-powered education.

Partnership development opportunities include relationships with schools, educational organizations, tutoring services, and other institutions that serve your target audience. These partnerships can provide user acquisition channels, credibility enhancement, and collaborative opportunities that accelerate platform growth.

Performance monitoring systems track user engagement metrics, agent effectiveness indicators, subscription conversion rates, user satisfaction scores, and other key performance indicators that inform ongoing platform development and marketing strategy refinement. These metrics provide the data needed for evidence-based improvement decisions.

User feedback collection mechanisms enable continuous platform improvement through systematic gathering of user suggestions, problem reports, feature requests, and satisfaction assessments. Educational platforms benefit particularly from regular user feedback because learning needs evolve continuously and platform effectiveness can always be enhanced.

Feature development planning establishes processes for prioritizing new capabilities, agent improvements, and platform enhancements based on user feedback, market opportunities, and technical feasibility. This planning should balance innovation with platform stability and user experience consistency.

Community building activities foster user engagement, peer learning, and platform advocacy through forums, user groups, success story sharing, and other mechanisms that create value beyond direct agent interactions. Educational communities can be particularly valuable because users often benefit from peer support and shared learning experiences.

Competitive analysis and market positioning ensure that your platform maintains distinctive value propositions while adapting to competitive pressures and market changes. The AI education market is evolving rapidly, requiring ongoing attention to competitive developments and strategic positioning refinement.

Financial management and growth planning address revenue optimization, cost management, profitability tracking, and investment decisions that support sustainable platform growth. Educational platforms often require patient capital and careful financial management during growth phases when user acquisition costs may initially exceed short-term revenue generation.

Long-term vision development guides strategic decisions about platform evolution, market expansion, additional agent development, and other opportunities that align with your overall business objectives while serving the educational needs of your growing user community.

Quality assurance and continuous improvement establish ongoing processes for maintaining agent effectiveness, platform reliability, user satisfaction, and educational outcomes that justify user investment in your platform. Educational technology requires continuous improvement because learning science advances and user expectations evolve constantly.

The successful implementation of these twelve chapters will transform your youraistudybuddy.com platform into a comprehensive AI-powered educational ecosystem that serves diverse learning needs while building a sustainable and growing business. Your unique combination of practical business experience, self-taught technical expertise, and deep commitment to educational excellence positions this platform for significant success in the rapidly expanding AI education market.

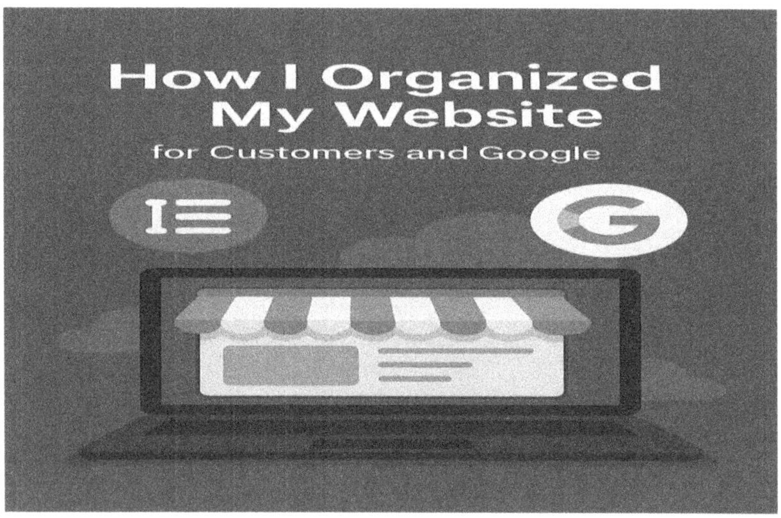

Chapter 34: How I Organized My Website for Customers and Google

One of the first lessons I learned while building **YourAIStudyBuddy.com** was this: if visitors can't find what they need in the first 10 seconds, they'll leave. That's why I didn't just focus on making my site look nice—I focused on making **it easy to navigate**.

The way you organize your site is just as important as the content itself. A well-structured website guides visitors like a friendly tour guide, while also sending clear signals to Google about what your site is all about.

Creating Core Pages

I started by listing out the essential pages every professional website should have. For me, that included:

1. **Home** – A welcoming overview with quick links to everything important.
2. **Shop** – Where customers could browse and buy products.
3. **About Us** – My personal story and the mission behind the site.
4. **Contact** – A simple way to get in touch.
5. **FAQ** – To answer common questions upfront.
6. **Blog** – A place for updates, tips, and helpful articles.

From the WordPress dashboard, I went to:
Pages → Add New, and started building each one.

At first, I kept things simple: a clear headline, a short paragraph, and maybe an image. Later, I went back and added special touches like testimonial blocks, product grids, or buttons leading to my subscription plans.

Using the Block Editor for Layout

The block editor turned each page into a blank canvas. With it, I was able to design layouts without hiring a developer.

- On my **About Us** page, I placed a photo of myself alongside my story using an *Image block* and a *Paragraph block.*
- On the **Contact** page, I used *WPForms* to insert a working contact form.
- On my **Shop** page, I added a *Product Grid block* (via WooCommerce) so items displayed neatly.

It was all drag, drop, and click—no code required.

Building Menus That Made Sense

Once I had my pages, I needed to connect them in a way that made sense. That's where menus came in.

From the dashboard I went to:
Appearance → Menus → Create New Menu.

I added my core pages, arranged them in the order I wanted, and assigned the menu as my **Primary Menu** so it appeared at the top of every page.

👉 One mistake I made early on was trying to include *everything* in the top menu. It made the site feel crowded and confusing. Once I trimmed it down to just 5–6 links, it instantly felt cleaner and easier to use.

Dropdowns and Categories

For bigger sections like **Shop** and **Blog**, I created dropdown menus.

- **Shop**
  – Books & Resources
  – Subscriptions
  – Special Offers
- **Blog**
  – Tutorials
  – Behind the Scenes
  – News

This way, visitors didn't have to dig around—they could jump straight to what they were looking for.

Making Navigation Mobile-Friendly

Most of my visitors use phones or tablets, so I tested my site on mobile constantly. Thankfully, most modern WordPress themes automatically turn menus into a mobile "hamburger" icon.

Still, I clicked through every page myself to make sure buttons weren't too small, text was readable, and the flow felt smooth on a smaller screen.

SEO Bonus: Naming Pages Smartly

What I didn't realize at first was that **page names help with SEO too**.

Instead of vague titles like *"Page 1"* or *"New Page"*, I used clear, descriptive names:

- **/learn-ai**
- **/ai-dictionary**
- **/downey-flower-shop** (for one of my other projects)

These cleaner URLs helped Google understand my site and made links look professional when shared.

Final Thoughts

Organizing my website turned out to be one of the most important things I did—both for my customers and for search engines.

Customers could now:

- Find what they needed in just a few clicks

- Navigate smoothly whether on desktop or mobile.
- Feel confident they were in the right place.

Google could now:

- Understand my site's structure.
- Index pages faster.
- Send me more of the right visitors.

I realized that a good site structure is like a well-designed storefront window: it invites people in, makes them feel comfortable, and encourages them to stay. And the best part? I didn't need to be a web developer—I just needed to think like a visitor and organize step by step.

Trust

Chapter 35: SEO (Search Engine Optimization)

How I'm Getting My Website Found on Google

When I launched my eCommerce site, I quickly realized that having a beautiful design wasn't enough — people also needed to **find** it. That's where SEO comes in. Search Engine Optimization helps your website appear higher in search results like Google, Bing, or Yahoo. The higher your site shows up, the more likely people are to click, visit, and eventually become customers.

Understanding Keywords

I started by thinking about what my audience — parents, students, and educators — would type into Google to find a business like mine. These are called **keywords**. For example:

- *"AI learning tools for kids"*
- *"easy AI dictionary"*
- *"AI videos for beginners"*

Once I had a list, I sprinkled these naturally into my page titles, product descriptions, blog posts, and even image file names. I didn't force them in — I just used them where they made sense.

Using an SEO Plugin

To make things easier, I installed an SEO plugin (like **Yoast SEO** or **All in One SEO**). These plugins work like a built-in coach, giving real-time feedback:

- Is your headline too long?
- Do you have a meta description?
- Are you using your keyword in the right places?

With those tips, I could adjust each page before hitting Publish.

Writing SEO-Friendly Content

Google rewards websites that actually **help people**. So instead of stuffing keywords, I wrote content that solved problems and answered questions clearly.

I also made sure to:

1. Use headings (H1, H2, H3) to structure pages.
2. Link to other pages on my site (internal links).
3. Keep paragraphs short and easy to read.

Image Optimization

Large images can slow down your site — and slow sites don't rank well. To fix this, I:

- Compressed images before uploading.
- Added descriptive alt text so search engines understood what each image was about. For example, instead of "image1.jpg," I used "ai-learning-book-for-kids.jpg."

Creating a Sitemap

My SEO plugin also generated a **sitemap** — a roadmap for search engines that shows how all my pages connect. I submitted it through **Google Search Console**, which helped Google index my site faster.

Local SEO (Bonus Tip)

If your business serves a local area, include your city or neighborhood in your titles and descriptions. For example:

- *"Downey AI Learning Tools for Kids"*
  That way, people searching nearby are more likely to find you.

Staying Updated

SEO isn't a "set it and forget it" thing. I check my analytics regularly to see which pages are doing well and which need improvement. If a page isn't getting much traffic, I'll try:

- Rewriting the headline
- Adding more helpful content

- Improving the images or alt text

Every little tweak adds up.

Final Thoughts

I know we've touched on some of these ideas before, but repetition is how things stick. The more natural SEO becomes, the easier it will be to grow your site long term

For me, SEO turned my site from "just another website" into one people can actually discover. And that's the real magic — not just building something beautiful, but making sure the right people find it.

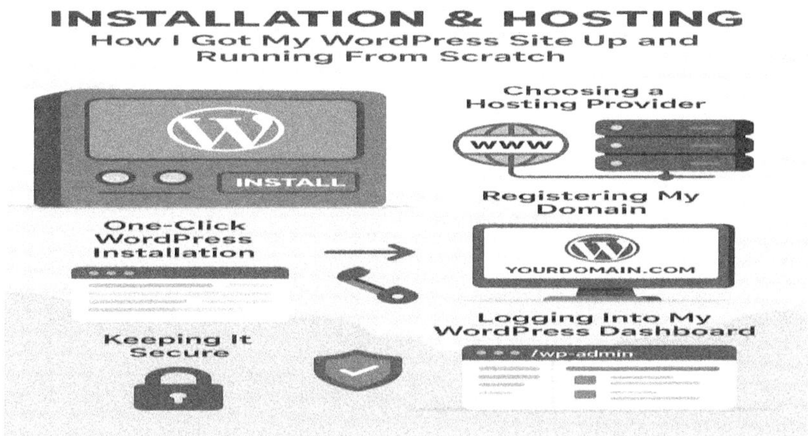

Chapter 36: Installation & Hosting – Getting Your WordPress Site Online with Hostinger

When I launched *YourAIStudyBuddy.com*, my first big goal was simple: make it live and ready for the world. That meant choosing a hosting provider, securing my domain, installing WordPress, and protecting the site from day one. At first, it sounded like something only a "techie" could do. But once I understood each step, it felt empowering—like I was building the foundation for my dream.

Step 1 – Choosing Hostinger as My Launchpad

A **hosting provider** is the company that stores your website's files and delivers them to visitors when they type in your domain. Without hosting, your site can't be seen.

I chose **Hostinger** because it gave me:

- ⚡ Fast loading speeds (great for users *and* SEO)
- 🔒 99.9% uptime so my site would always be online
- 💬 24/7 live chat support for quick help
- 💡 One-click WordPress installation

- 💸 Affordable plans that could grow with my business

For me, Hostinger wasn't just storage—it was the launchpad that turned my idea into a real, live website.

Step 2 – Securing My Domain Name

Your **domain name** is your website's unique address on the internet, like *YourAIStudyBuddy.com*. It's how people find you, and it's part of your brand identity.

I already owned my domain, but if I hadn't, I could have registered it directly through Hostinger during sign-up. My tip? Pick something clear, memorable, and professional—it makes you easier to find and builds instant trust.

Step 3 – Installing WordPress with One Click

WordPress is the platform that powers your website—the tool that lets you design, publish, and manage everything without coding.

With Hostinger, installation was as easy as:

1. Logging into my account
2. Going to **Website → Auto Installer**
3. Selecting WordPress
4. Entering my site title, username, and password
5. Clicking **Install**

In minutes, my WordPress site was ready for customization. No stress, no confusion—just a clean start.

Step 4 – Connecting My Domain

If your domain is registered somewhere else, you'll need to "point" it to your hosting provider. This is done by updating your **DNS settings** to Hostinger's nameservers.

I logged into my domain registrar, updated the nameservers provided by Hostinger, saved the changes, and within a few hours, my domain was live and connected. Think of it like updating your phone number in the address book—so when people look for you, they reach the right place.

Step 5 – Accessing the WordPress Dashboard

To start building, I went to:

👉 **www.YourAIStudyBuddy.com/wp-admin**

I logged in with my admin username and password, and there it was—the **WordPress Dashboard**, my control center. From here I could:

- Create pages and posts
- Install and activate themes
- Add plugins for new features
- Manage security, design, and settings

This was the moment my blank canvas turned into the home for my vision.

Step 6 – Securing My Site From Day One

Building a website is exciting, but protecting it is essential. From the start, I:

- Installed a trusted **security plugin**
- Created a strong, unique password
- Enabled **two-factor authentication**
- Set up daily automatic backups

Why? Because a secure website builds trust with your visitors and saves you from painful disasters later. Prevention is far easier than repair.

Final Note

Installing WordPress with Hostinger was far less intimidating than I imagined. In under an hour, I went from *"idea"* to *"live website."*

The best part? This wasn't just a technical step—it was a foundation. A solid starting point for everything that came next: building pages, adding products, connecting Stripe, and launching an interactive AI learning hub.

If you're starting your own site, remember this: **you don't need to be a tech expert. You just need a clear process and the courage to start.**

Chapter 37: Polishing the Site – Menus, Pages, and Final Touches

With the foundation in place, WordPress installed, and content added, I shifted my focus to the **user experience**. *YourAIStudyBuddy.com* was filled with books, videos, subscription plans, and an AI chatbot—but none of that would matter if visitors couldn't navigate smoothly.

This chapter is about transforming a functional website into an inviting, professional, and easy-to-use learning space.

Step 1 – Structuring the Main Menu

The main menu is the navigation bar at the top of your website. It's the "map" visitors rely on. A clear, well-organized menu reduces confusion and keeps people exploring.

My main menu included six purposeful tabs:

1. Home – Welcoming introduction

2. Learn AI – Lessons and videos in one place
3. Books & Downloads – IntelliGloss, AI books, and resources
4. Choose a Plan – Subscription tiers with signup links
5. Login / My Account – Secure access for returning users
6. Chat With Your AI Buddy – Direct link to the chatbot

Every tab was chosen to guide, engage, and convert visitors.

Step 2 – Creating Clean, Focused Pages

Each page is like a room in your digital house. Too much clutter overwhelms visitors; clean layouts keep them engaged.

On *YourAIStudyBuddy.com*, I designed pages such as:

- **Choose a Plan** – Four subscription tiers in colorful boxes with bold "Subscribe" buttons.
- **Books & Downloads** – Cover images, previews, and pricing.
- **Learn AI** – Video thumbnails with simple "Play" buttons.

Every section flowed naturally into the next, making scrolling feel effortless.

Step 3 – Branding With Visuals

Branding builds trust and recognition. Using WordPress's Site Editor, I set global colors, fonts, and button styles.

Here's what I added:

- Brand colors (blues and purples) for headers and buttons
- Book cover images and educational icons
- My logo and tagline — *"Your AI Study Buddy – Where Learning Meets the Future"* — displayed on every page

Step 4 – Testing Mobile Responsiveness

More than half of internet traffic comes from phones. I previewed my site on multiple devices and adjusted spacing, button sizes, and text for easy reading and tapping.

Even the chatbot was tested to ensure it opened smoothly on small screens.

Step 5 – Double-Checking Links and Speed

Broken links and slow load times frustrate visitors and hurt SEO rankings.

- Tested every button and menu link
- Removed unused drafts and outdated pages
- Compressed large images
- Cleared my site cache to keep performance sharp

Polishing Checklist ✅

Before launch, make sure you:

- Menus are simple, clear, and organized
- Each page has one clear purpose
- Branding (colors, fonts, logo) is consistent
- Site looks good on desktop, tablet, and mobile
- All links and buttons work
- Images are optimized and site speed is tested

Final Thoughts

Building a website is one thing—**polishing it is another**. These finishing touches turned *YourAIStudyBuddy.com* from a working platform into a complete learning experience.

Now when someone lands on my homepage, they don't just see content—they feel welcomed, guided, and supported. That's the difference between a site that functions and a site that inspires.

Chapter 38: What I Learned – Reflections, Wins, and What's Next

Building *YourAIStudyBuddy.com* wasn't just a technical project — it was a personal journey. I learned how to install WordPress, connect Stripe, and design a chatbot, yes. But more than that, I learned about resilience, creativity, and the power of starting with nothing but a dream and the courage to try.

Here are the biggest lessons I'm taking with me — and what's next.

1. The Power of Starting Small

I didn't have a development team or a big budget. What I did have was:

- A notebook full of ideas
- A passion for education
- A willingness to learn

I began with the basics: hosting, WordPress, a theme, and a few pages. Mistakes happened, but each one taught me something new. By starting small, I was able to grow with confidence.

Today, I can proudly say I built a functional eCommerce site, integrated a chatbot, and launched a global AI learning platform — on my own.

2. Celebrate Every Milestone

Progress often came in small steps, but I celebrated each one:

- My first Stripe payment
- My first chatbot conversation

- The first teacher who asked for IntelliGloss
- The first parent who said, "My child loves your book"

Each milestone fueled my motivation and reminded me why I started.

3. Tech Isn't Just for the Tech-Savvy

I used to believe building websites and AI tools was only for coders. Now I know better. Tools like WordPress, Hostinger, and no-code chatbots make the digital world accessible.

If I can do it, with a background in floral design and education, then so can you.

4. Community Is Everything

No project is truly "solo." My community was made up of:

- YouTube tutorials
- Online forums and guides
- My AI assistant
- Friends and family cheering me on

Now, I want *YourAIStudyBuddy.com* to become that kind of community — a place where students, parents, and educators can learn, ask, and grow together.

5. What's Next

This isn't the end — it's the foundation. Here's what I'm working on:

- Publishing more children's AI books
- Creating video courses for teens and beginners
- Adding interactive quizzes and learning tools
- Expanding IntelliGloss with more AI terms
- Writing my next guidebook: *How to Build Chatbots and eCommerce Websites Without a Tech Background*

Each step serves the same mission: making AI learning accessible, friendly, and inspiring.

Quick Recap of My Wins 🎉

- Built and launched *YourAIStudyBuddy.com*
- Integrated Stripe payments and created subscription plans
- Designed an interactive AI chatbot with Botpress
- Published IntelliGloss and children's AI books
- Reached students, parents, and teachers worldwide

Final Words

YourAIStudyBuddy.com isn't just a website — it's proof that with faith, focus, and determination, anyone can create something meaningful.

To everyone reading this:
Keep learning.
Keep building.
Keep believing in your ideas.

Because with the right mindset and a little guidance, you can create something that changes the world — or at least someone's world — for the better.

Chapter 39: Adding a Chatbot to Your WordPress Site – My Step-by-Step Botpress Journey

When I first started building *YourAIStudyBuddy.com*, I didn't want it to just be a place to buy books or watch videos. I wanted it to feel interactive — like a real study buddy was there to guide students, answer questions, and support learners.

That's when I discovered **Botpress**, an open-source chatbot builder powered by AI. What I lacked in technical skills, I made up for with curiosity and persistence.

What Is Botpress and Why I Chose It

Botpress is an open-source conversational AI platform that lets you build customizable chatbots without coding everything from scratch.

Here's why I chose it:
✅ Natural language understanding (NLU) to make the bot smarter
✅ Ability to train it with my own content
✅ Easy WordPress integration
✅ Affordable and open source

It felt like hiring a 24/7 support rep for my website — one that never sleeps!

Step 1: Signing Up

1. Go to www.botpress.com
2. Create a free account
3. Verify your email
4. Create a workspace (I named mine *Your AI Study Buddy*)

Now you're ready to build your first bot.

Step 2: Building the Chatbot

Inside Botpress Studio, I:

- Created a friendly welcome message
- Built flows to answer FAQs like *"What is AI?"* or *"Where can I buy your book?"*
- Added quick-reply buttons and fallback replies
- Trained the bot to recognize phrases like *"learn AI,"* *"study plans,"* and *"free trial."*

It felt more like arranging building blocks than writing code.

Step 3: Embedding the Chatbot in WordPress

Here's how I made the bot appear on my site:

1. In Botpress, I copied the **embed code** (found under Deploy → Webchat → Embed Code).
2. In WordPress, I went to **Appearance → Theme File Editor → footer.php**
3. Pasted the code right before the `</body>` tag
4. Saved changes — and refreshed my site.

In seconds, my chatbot popped up in the corner, ready to help!

👉 *Pro Tip: If you don't want to edit code, use the free "Insert Headers and Footers" plugin. Just paste the code there and you're done — no file editing required.*

Step 4: Customizing the Look and Feel

Botpress let me:

- Change the chat bubble color
- Pick the position (left or right corner)
- Adjust the welcome message tone

I matched everything to my brand: simple, warm, and student-friendly.

Step 5: How It's Helping Users

Since launching the chatbot, visitors are:

- Getting instant answers
- Finding plans and products faster
- Feeling more confident navigating the site

This reduced support questions — and increased engagement.

Lessons Learned

1. Be patient. The first version won't be perfect.
2. Train the bot with *real questions* from users.
3. Keep refining. A chatbot is always learning.

Final Thoughts

Adding a chatbot to my WordPress site wasn't just a tech upgrade — it was a personal achievement. Now when someone lands on *YourAIStudyBuddy.com*, they're greeted by a smart, helpful AI assistant that's always available.

And the best part?
I built it. And so can you.

Chapter 40: Adding a Chatbot to Your WordPress Site – My Step-by-Step Botpress Journey

When I built *YourAIStudyBuddy.com*, I didn't want it to be just another eCommerce site. I wanted it to feel alive and interactive — like a real study buddy was guiding students, answering questions, and offering support 24/7.

That's when I discovered **Botpress**, an open-source AI chatbot platform. I didn't need coding skills — just curiosity, persistence, and a clear vision.

What Is Botpress and Why I Chose It

Botpress is a customizable chatbot builder powered by Natural Language Understanding (NLU). It interprets user questions, lets you train it with your own content, and integrates seamlessly with WordPress.

I chose it because it was:

- Open source and affordable
- Easy to embed into WordPress
- Flexible enough to grow with my site
- Designed with a visual, no-code dashboard

For me, it felt like hiring a round-the-clock support rep who never takes a coffee break.

Step 1: Signing Up

1. Visit **botpress.com**
2. Create a free account and verify your email
3. Name your workspace (I called mine *Your AI Study Buddy*)

Step 2: Building the Chatbot

Inside Botpress Studio, I:

- Created a friendly welcome message
- Added flows for FAQs like *"What is AI?"*
- Built quick-reply buttons and fallback responses
- Trained it with phrases like *"learn AI," "study plans,"* and *"free trial"*

Step 3: Connecting to WordPress

1. In Botpress, copy the embed code (Deploy → Webchat → Embed Code).
2. In WordPress, paste it into **footer.php** just before the `</body>` tag.
3. Save and refresh — the chatbot appears instantly.

☞ *Tip: If coding feels intimidating, use the free Insert Headers and Footers plugin to paste the code safely.*

Step 4: Customizing the Look

I matched my chatbot to *YourAIStudyBuddy.com* by adjusting:

- Bubble color
- Position (bottom right)
- Greeting tone
- Avatar image

Step 5: How It Helps Visitors

Since adding the bot, visitors can:

- Get instant answers
- Find the right plan or product faster
- Navigate the site with confidence

It reduced support questions — and increased engagement.

Lessons Learned

- Be patient: the first version won't be perfect

- Train it with real user questions
- Keep refining — a chatbot is never "done"

Final Thoughts

Adding a chatbot wasn't just a technical upgrade — it was a personal achievement. Now, every visitor to *YourAIStudyBuddy.com* is greeted by a smart, friendly AI assistant.

And the best part? **I built it — and so can you.**

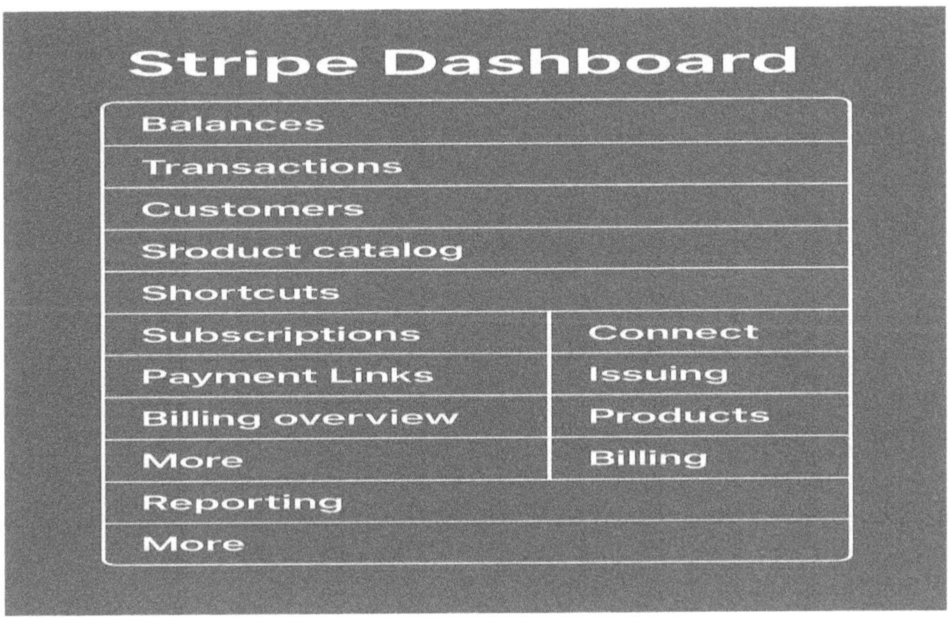

Chapter 41: Setting Up Stripe – Accepting Payments Worldwide on YourAIStudyBuddy.com

Getting my website ready to accept payments was the turning point that transformed YourAIStudyBuddy.com from a personal project into a real business. I needed a payment system that was secure, easy to use, and trusted by customers across the world.

That's why I chose **Stripe**. It's fast to set up, integrates seamlessly with WordPress, and works in multiple countries. Most importantly, it allows parents, students, and teachers to pay in ways they already trust—whether that's by credit card, debit card, or digital wallet.

In this chapter, I'll walk you through exactly how I connected Stripe, tested it, and celebrated my very first payment notification.

1. Why I Chose Stripe

Before deciding, I looked at PayPal, Square, and a few others. Stripe won me over for three key reasons:

- **Global reach** – My audience isn't just in the U.S. Stripe accepts payments from many countries and automatically supports multiple currencies, so learners in Fiji, Tonga, New Zealand, and beyond could check out without hassle.
- **Seamless WordPress integration** – It connects directly with WooCommerce and plugins like WP Simple Pay, so no complicated coding was required.
- **Automated payouts** – Funds go straight into my bank account without me needing to manually transfer anything.

For me, Stripe wasn't just a payment processor—it was the bridge that connected my work to customers worldwide.

2. Set Up Your Stripe Account

Getting started was simple:

1. I went to **stripe.com** and clicked **Get Started**.
2. Entered my business details (name, address, tax info, and bank account).
3. Verified my email.

Within minutes, my account was approved and ready to use.

3. Connect Stripe to WooCommerce

Once my Stripe account was ready, I integrated it into WordPress:

1. From my WordPress dashboard, I went to **Plugins > Add New**.
2. Searched for **WooCommerce Stripe Payment Gateway**, then installed and activated it.
3. Navigated to **WooCommerce > Settings > Payments**, enabled Stripe, and clicked **Manage**.
4. Entered the **API keys** from my Stripe dashboard (both test and live keys).
5. Saved my settings.

Just like that, Stripe was live and ready to process payments.

4. Test Your First Payment

Before going live, I wanted to be sure everything worked. Stripe provides "test card numbers" so you can simulate purchases without charging real money.

Here's what I did:

- Placed a $1 test order for a sample product.
- Checked that WooCommerce showed the payment as **completed**.
- Verified the transaction inside my Stripe dashboard.

When everything worked perfectly, I switched from **Test Mode** to **Live Mode**—ready for real customers.

5. Add Flexible Payment Options

I wanted every visitor to feel comfortable checking out, so I enabled multiple options through Stripe:

- Credit and debit cards
- Apple Pay and Google Pay
- Local currency support (so customers abroad see prices in their own currency)

This flexibility meant that a parent in Australia, a student in Fiji, or a teacher in the UK could all subscribe without a second thought.

6. The Moment It Became Real

I'll never forget the first email notification from Stripe:

"You've received a payment."

It wasn't just about money. It was proof that someone believed in my work enough to invest in it. That moment confirmed that all the late nights learning, troubleshooting, and building were worth it.

7. What Came Next

With Stripe in place, the store was officially open for business. Now I could sell:

- **IntelliGloss**, my AI dictionary for students
- My **children's AI books**
- **Subscription plans** with videos and lessons

The payment system was secure, the chatbot was in place, and my site was finally ready to serve learners around the globe.

✅ **Final Thoughts**
Stripe was the missing piece that transformed YourAIStudyBuddy.com into more than just a website—it became a business. Setting it up was simpler than I expected, and it gave me the confidence to know that when someone clicks "Buy," everything works seamlessly behind the scenes.

In the next chapter, I'll share how I organized my books, videos, and downloadables into a smooth learning hub—so visitors could not only pay easily, but also find exactly what they needed right away.

Chapter 42: Organizing and Selling Digital Products – Books, Videos, and Lessons on YourAIStudyBuddy.com

Once the payment system was set up and the chatbot was in place, the next big step was organizing my products — especially my AI books, video lessons, and downloadable study materials.

Here's how I turned YourAIStudyBuddy.com into a structured learning hub where students, parents, and teachers could easily browse, preview, and purchase content.

1. Creating a Student-Friendly Catalog

I started by thinking like a visitor. If I were new to the site, how would I want to explore? The answer was simple categories:

- **AI Books** – Children's books, study guides, and future eBooks
- **Video Lessons** – Bite-sized videos like *What is AI?* or *How Does a Chatbot Work?*
- **Downloadables** – Worksheets, quizzes, notes, and infographics

Each category had its own page, with clean previews and clear purchase options.

2. Selling Digital Books

WooCommerce made selling eBooks easy:

1. Went to **Products → Add New**
2. Uploaded the book cover, wrote a short description, and set the price
3. Checked **Downloadable** and **Virtual** so customers got an instant link after purchase
4. Grouped them under the category **Books & eBooks**

Now, every book appeared neatly with cover art, pricing, and a one-click checkout.

3. Adding and Previewing Video Lessons

Videos are one of the most engaging ways to learn, but I wanted to balance free previews with subscriber-only access.

Here's what I did:

- Uploaded **short preview clips** to YouTube (Unlisted) and embedded them on the site
- Hosted **full lessons** privately and unlocked them for logged-in subscribers
- Organized lessons into categories like:
    - Intro to AI
    - Intermediate Learning
    - Advanced Thinking

This gave learners a taste of the content before committing — and encouraged upgrades.

4. Uploading Downloadable Resources

Printable worksheets and guides became a favorite feature for parents and teachers.

- Uploaded PDFs into the **WordPress Media Library**
- Linked them in WooCommerce using the **Downloadable Files** option
- For freebies, I added a simple button using the **Simple Download Monitor** plugin

Everything was labeled clearly so visitors knew exactly what they were getting.

5. Building the AI Learning Hub

To pull everything together, I created a central **Learn AI** page. It included:

- Quick links to Books, Videos, and Downloads
- A **"Start Here"** button for beginners
- My chatbot, offering to guide users based on their level
- A free trial sign-up prompt

This page became the heart of the site — a launchpad for every learner.

6. Connecting Products to Subscription Plans

To keep the experience seamless, I mapped products to my subscription tiers using a membership plugin:

- **Free Trial** – Previews and a few downloads
- **Standard Plan** – Beginner lessons + one eBook
- **Premium Plan** – All books, all videos, and all downloads
- **Elite Plan** – Everything above + live Q&A and chatbot support

This system made it clear what each plan unlocked — no confusion, no frustration.

Final Thoughts

The secret to selling digital products online isn't having the most content — it's organizing it in a way that feels natural and simple for your audience.

By creating categories, using previews, and linking products to subscription levels, I turned YourAIStudyBuddy.com into more than a store. It became a guided learning experience where every visitor, no matter their age or skill, could find a starting point and grow at their own pace.

.Chapter 43: Setting Up Automated Emails — Welcoming and Engaging Your Users

When YourAIStudyBuddy.com began attracting its first users, I quickly realized something important:

Having a great website and quality content was only half the battle. If I wanted people to *stay*, explore, and eventually subscribe, I needed a way to keep in touch — automatically.

That's where **email automation** became my secret weapon. It welcomed new users, guided them through the site, reminded them of trial deadlines, and encouraged upgrades — all without me sending a single manual email.

Think of it as having a personal assistant working 24/7, making sure every learner feels supported.

1. Why Email Automation Matters

Every sign-up is a potential long-term learner — but many drift away without guidance. I asked myself:

"If I could sit down with each new user, what would I say to help them succeed?"

That question shaped my goals:

- **Welcome warmly** — First impressions build trust.
- **Guide them** — Show where to start and how to get value.
- **Remind them** — Keep trials and deadlines visible.
- **Encourage upgrades** — Offer the next step when they're ready.
- **Re-engage the quiet ones** — A gentle reminder can reignite interest.

Automation made this possible without extra daily work.

2. Choosing the Right Plugin

I explored several WordPress-friendly options:

- **MailPoet** — beginner-friendly and built into WordPress
- **FluentCRM** — powerful automation and segmentation
- **Mailchimp for WooCommerce** — widely used, flexible
- **WP Fusion** — deep integrations for membership sites

I chose a plugin that let me trigger emails based on actions like signing up, buying a book, or going inactive.

3. Crafting a 3-Part Welcome Series

Your welcome sequence is like rolling out a red carpet. Mine looked like this:

1. **"Welcome to Your AI Study Buddy 🎉"**
   Sent immediately. Introduced the platform, outlined plan benefits, and linked to the dashboard.
2. **"How to Get the Most from Your Plan"**
   Sent 24 hours later. A guided tour pointing to videos, downloads, and the chatbot.
3. **"Meet Your Learning Buddy 🤖"**
   Sent 3 days later. Explained how my Botpress chatbot could answer AI questions and suggest resources.

Each email was short, friendly, and written like I was talking to a friend.

4. Trial Reminders & Upgrade Invitations

For free trial users, I set up:

- **Day 25:** "Your Free Trial Ends Soon" — framed as a helpful reminder.
- **Day 30:** "Let's Keep Learning Together" — with a small incentive (like 10% off).

Not pushy — just supportive nudges to continue the journey.

5. Transactional Emails That Build Trust

Automated confirmations also mattered:

- **Order confirmation**

- **Receipt & subscription start**
- **Thank you email** linking directly to purchased content

I branded each one with my logo, colors, and tone so they felt professional and reassuring.

6. Re-Engagement Emails

For users who went quiet, I set up a gentle reminder:

Subject line: "We Miss You at Your AI Study Buddy 💡"
Inside, I shared a popular video or resource and invited them back. Many returned.

7. Respecting Privacy & Legal Compliance

Every email included:

- A clear **unsubscribe link**
- My **business address**
- A short note on **data privacy**

This kept me compliant with GDPR and CAN-SPAM — and, more importantly, showed respect for my users.

Final Thoughts

Automated emails turned YourAIStudyBuddy.com into more than a website — they created an ongoing conversation. From welcome sequences to re-engagement nudges, users felt guided and supported every step of the way.

The best part? I didn't need a marketing team. Just the right plugin, a clear plan, and a little creativity.

With automation in place, I wasn't just gaining subscribers — I was building a learning community.

Chapter 44: Managing Members, Creating Subscription Plans, and Protecting Premium Content

As YourAIStudyBuddy.com grew, I needed a way to organize members at different levels — from free trial learners to elite subscribers — while protecting premium content so only paying users could unlock it.

This chapter shows how I structured subscription plans, connected them with Stripe, and used WordPress tools to make the whole system seamless.

1. Defining My Subscription Levels

I started by mapping out what each membership tier should include. Having clear levels made it easy for users to understand their options and for me to manage access.

Plan	Price	What's Included
Free Trial (30 days)	Free	Limited chatbot use, 3 free videos, 1 downloadable worksheet
Standard	$10/month	All basic videos, all worksheets, limited chatbot access
Premium	$19.99/month	Everything in Standard + full video library, AI glossary book, custom learning paths
Elite	$29.99/month	Everything unlocked + 1-on-1 coaching, live Q&A, full chatbot access

This tiered structure not only organized my site but also gave users a clear upgrade path.

2. Installing a Membership Plugin

To manage subscriptions and content restriction, I explored plugins like:

- Paid Memberships Pro
- Restrict Content Pro
- MemberPress

- WP Simple Pay (for simpler setups)

I chose the one that offered Stripe integration, flexible membership levels, and easy plan upgrades/downgrades.

3. Connecting Stripe for Recurring Payments

Stripe was already set up for one-time purchases, but subscriptions required recurring billing.

Inside the plugin's settings, I:

- Created matching subscription plans in Stripe
- Linked them to WordPress membership tiers
- Enabled auto-renewal and easy cancellations
- Customized the checkout flow

The result: members signed up, paid, and instantly gained the right access — no manual work from me.

4. Protecting Premium Content

Next, I set rules so content unlocked only for the right users:

- 🔒 **Videos** – Restricted with shortcode blocks by plan level
- 🔒 **PDFs & eBooks** – Download buttons hidden unless logged in with the right plan
- 🔒 **Chatbot** – Limited for trial users, full access for Elite

This created a smooth but secure experience.

5. The "My Account" Dashboard

A self-service account page gave users the ability to:

- View their current plan
- Upgrade or cancel subscriptions
- Manage payment details
- See past invoices and downloads

This empowered members and reduced the number of support emails I received.

6. Handling Expired Plans Automatically

When a user didn't renew, the system:

1. Downgraded them to Free Trial
2. Locked their premium content
3. Sent a friendly email inviting them back

This automation ensured no one got frustrated — everything happened smoothly in the background.

7. Offering Discounts and Promos

To encourage signups and upgrades, I added:

- One-time coupon codes
- Seasonal discounts
- Trial extensions for loyal users

This flexibility gave me more ways to connect with my audience and reward engagement.

Final Thoughts

By structuring clear subscription plans, connecting them to Stripe, and using WordPress tools to manage access, I turned YourAIStudyBuddy.com into a true membership-based learning platform.

The system gave me peace of mind and gave users confidence. Whether they were exploring the free trial or joining the Elite plan, their experience felt smooth, professional, and valuable.

If you're just starting, remember this: you don't need a tech background to build a membership site. All you need is a clear vision of the value you want to offer — and the right tools to bring it to life.

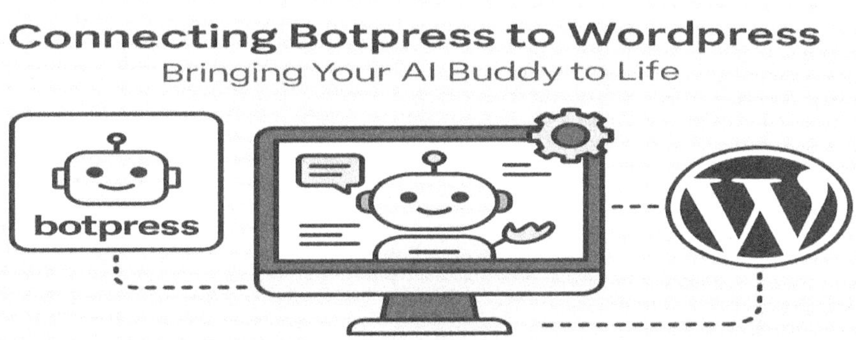

Chapter 45 — Taking Your Chatbot Further with Botpress

When I first added a chatbot to YourAIStudyBuddy.com (see Chapters 28–29), I thought I was done. The bot could greet visitors, answer simple questions, and guide people to my books and lessons. But over time, I realized that a chatbot isn't just a "set it and forget it" feature — it's a tool that grows with your users.

In this chapter, I'll share the advanced steps I took to transform my chatbot from a basic helper into a true digital assistant for my students and subscribers.

1. Analyzing Real Conversations

What it is:
Botpress comes with analytics that show you what users are asking, where conversations break down, and which responses work best.

How I used it:

- I looked for repeated questions my bot couldn't answer well.
- I added new responses based on what people were actually asking (like "Can I print the worksheets?").
- I refined my welcome message when I noticed users often dropped off early.

Why it matters:
Instead of guessing, I used real data to make my bot smarter.

2. Training with FAQs and My Own Content

What it is:
Botpress lets you feed the chatbot your own documents, FAQs, or content so it learns from *your knowledge*.

How I used it:

- I uploaded my **AI glossary** so the bot could explain terms directly.
- I added **customer FAQs** (e.g., "How long is the free trial?" or "Do you ship internationally?").
- I kept responses short, clear, and friendly.

Why it matters:
The more context your chatbot has, the more it feels like a real study buddy.

3. Adding Multilingual Support

What it is:
Botpress can be trained to understand and reply in multiple languages.

How I used it:
Since my readers come from Fiji, Tonga, New Zealand, and the U.S., I trained the chatbot with English first, then added key phrases in other languages my audience speaks.

Why it matters:
A simple "Bula!" or "Kia ora!" greeting makes international visitors feel instantly welcome.

4. Integrating with WooCommerce

What it is:
With a little setup, your chatbot can connect to your store to suggest products or guide users through checkout.

How I used it:

- Added quick replies like **"Show me AI books"** or **"Subscribe now"**.
- Linked bot responses directly to product pages.
- Tested the flow so users could move from chatbot → product → checkout in just a few clicks.

Why it matters:
Instead of being "just a helper," the chatbot became a sales assistant.

5. Personalizing the Experience

What it is:
Botpress lets you recognize returning users and customize messages.

How I used it:

- Subscribers see: *"Welcome back, [Name]! Ready for your next lesson?"*
- Free trial users see: *"Hi! Did you know your trial ends in 5 days? Here's how to upgrade."*
- Parents who bought a book see: *"Want the matching worksheets? Click here to download."*

Why it matters:
Personalization makes users feel like the chatbot *knows them* instead of giving canned answers.

Final Thoughts

A chatbot isn't just a gadget — it's an ongoing project. The more I updated and improved mine, the more it became part of the YourAIStudyBuddy.com experience.

Now, instead of answering the same questions over and over, I let the chatbot handle it. Instead of losing visitors who couldn't find what they wanted, I guide them instantly. And instead of being "just a website," my platform feels alive — like a real buddy is always there to help.

Chapter 46: Launching YourAIStudyBuddy.com and Growing the Platform

Launching YourAIStudyBuddy.com was more than publishing a website — it was turning a dream into reality. My goal was simple: create a space where students, beginners, and lifelong learners could explore AI in a fun, simple, and meaningful way.

In this chapter, I'll share how I prepared for launch, welcomed new users, promoted the platform, and continue to keep it growing.

1. Preparing for Launch

Before telling the world, I made sure every detail was in place. I:

- Uploaded all books, videos, and downloads
- Tested Stripe in both test and live modes
- Checked membership plans (Free, Standard, Premium, Elite)
- Verified the chatbot's responses
- Cleaned up pages, menus, and broken links
- Optimized for mobile and fast loading
- Tested confirmation emails for signups and purchases

This preparation meant early visitors had a smooth, professional first experience.

2. Creating a Welcoming Onboarding Flow

First impressions are everything. To make newcomers feel at home, I added:

- A short welcome video introducing the platform
- A **Start Here** guide to show first steps
- A friendly chatbot opener: *"Want to learn AI the fun way?"*
- A free downloadable worksheet to encourage signups

This combination reduced confusion and made the platform feel approachable from day one.

3. Marketing the Launch

When launch day came, I spread the word across multiple channels:

- Email announcements to my contacts
- Facebook posts with book covers and pricing
- Sharing links on Amazon, Barnes & Noble, and education forums
- Short YouTube teasers with fun AI facts
- Cross-promotion through my other businesses

The focus wasn't on "selling" — it was on getting people to try the free trial, meet the chatbot, and see the value for themselves.

4. Gathering Feedback and Reviews

Early users shaped the platform with their feedback. I asked three simple questions:

1. What did you enjoy most?
2. What could we improve?
3. Would you recommend this to a student or school?

Their answers led me to refine layouts, simplify instructions, and improve chatbot responses. I also encouraged book and site reviews, which helped build trust with new visitors.

5. Partnering with Schools and Educators

To reach beyond individuals, I connected with:

- Teachers looking for fresh AI resources
- School librarians adding to STEM collections
- STEM programs and summer camps
- PTA groups supporting learning initiatives

I even created a printable brochure and sample kit for educators to share with decision-makers.

6. Keeping the Platform Fresh

A launch is just the beginning. To keep momentum, I:

- Add new books, worksheets, and short videos regularly
- Post content on trending AI topics
- Update plugins and themes for security and performance

- Review analytics to see what users engage with most

This cycle ensures learners always have something new to explore.

From Dream to Reality

Launching YourAIStudyBuddy.com proved that you don't need to be a tech expert to create something meaningful. With WordPress, Stripe, Botpress — and a clear mission — I built a platform that helps people learn and grow.

And this is only the beginning.

If you're reading this, I hope my story inspires you to start your own project — whether it's a course, a community, or a resource hub. With patience, creativity, and the right tools, you can launch something that matters to the world.

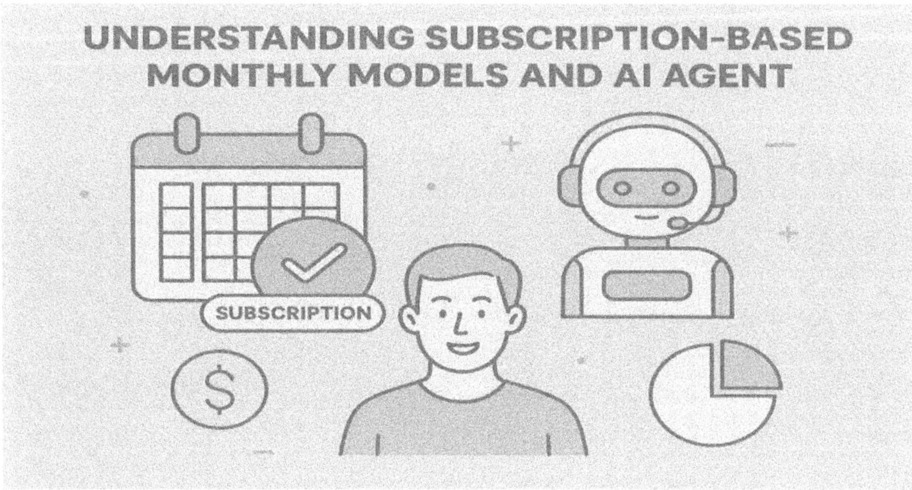

Chapter 47: Subscription Models and AI Agents — A Powerful Combination

When I launched YourAIStudyBuddy.com, I quickly learned something important: **a great website needs both a sustainable business model and a smart way to serve users.** That's where subscription plans and AI agents come together.

What Is a Subscription-Based Monthly Model?

A subscription model means customers pay a recurring fee — usually monthly — to access products, services, or content. Instead of making a one-time purchase, they enjoy **ongoing access and updates** for as long as they stay subscribed.

On YourAIStudyBuddy.com, this includes:

- Exclusive IntelliGloss digital content

- Premium AI chatbot features
- Full access to video lessons and downloadable worksheets

This model gives me predictable income, while learners get continuous value at an affordable cost.

What Is an AI Agent?

An AI agent is software powered by artificial intelligence that can perform tasks autonomously. On my site, it's the **AI Study Buddy chatbot** built with Botpress.

It can:

- Answer questions in natural language
- Guide users through the site
- Recommend learning resources
- Help with subscription decisions

Think of it as a 24/7 assistant who's always ready to help learners.

Why They Work So Well Together

While a subscription model is about revenue, and an AI agent is about technology, they reinforce each other beautifully:

- **The subscription** ensures steady income.
- **The AI agent** improves customer experience, making sign-ups, upgrades, and renewals easier.

Together, they attract new learners, keep them engaged, and encourage long-term loyalty.

How It Works in Practice

1. **Attracting Subscribers**
   One of my first trial visitors typed: *"What do I get with the Standard Plan?"* The chatbot instantly listed the benefits and linked them straight to the checkout page. That interaction turned into my very first subscription sale.
2. **Enhancing User Experience**
   Once subscribed, the AI agent provides personalized guidance, answers FAQs, and keeps learners on track.
3. **Boosting Retention and Upgrades**
   The bot can remind users about renewals or introduce higher-tier plans, encouraging them to stay longer.
4. **Providing Insights**
   Every interaction generates data. By analyzing what people ask most, I can refine both the chatbot and the subscription offers.

The Takeaway

A subscription model gives your business stability. An AI agent gives your users support. Together, they create a system that's **profitable for you, valuable for your customers, and sustainable for the long term.**

For me, the moment that first user subscribed — guided by my chatbot — proved that this approach really works. It was the clearest sign yet that YourAIStudyBuddy.com wasn't just an idea anymore. It was alive, growing, and ready to support learners around the world.

Chapter 48: Essential E-commerce Tools That Made My Site Shine

Building *YourAIStudyBuddy.com* to sell *IntelliGloss: The Definitive AI Dictionary* taught me one crucial lesson: the right tools turn a good idea into a professional, user-friendly business. As someone who didn't grow up coding—or navigating WordPress's endless menus and tricky hosting settings—I know how overwhelming it can feel. That's why I want to share the tools I chose for payments, domains, and hosting, so you can build your store with confidence.

Payments That Work Worldwide

To sell online, you need a gateway that handles payments securely and conveniently. I chose **Stripe**, and here's why:

- Supports 135+ currencies
- Accepts credit cards, debit cards, Apple Pay, and Google Pay
- Integrates seamlessly with WordPress and WooCommerce

Thanks to Stripe, I've been able to sell to customers in South Korea, Tonga, and the Netherlands—without extra setup.

Other great options:

- **PayPal** — A trusted classic, especially for buyers who prefer one-click checkouts.
- **Square** — Excellent if you plan to sell in person as well as online.
- **Adyen, 2Checkout, and Worldpay** — Good for localized payment options (like iDEAL in the Netherlands), boosting conversions in specific regions.
- **Braintree** — Stripe's closest competitor, known for global scalability and flexibility.

Choosing the Right Hosting & Site Builder

I built my site with **WordPress** hosted on **Hostinger**. While WordPress gave me flexibility, I found Hostinger's interface a bit confusing at first. What saved me was pairing it with **Elementor**, a visual page builder that made designing pages feel as easy as arranging slides.

If you'd prefer an all-in-one platform without plugins or theme setup, alternatives include:

- **Shopify** — Tailor-made for e-commerce with built-in tools.
- **Wix** — Beginner-friendly with drag-and-drop editing.
- **Squarespace** — Great for professional-looking designs with minimal setup.

Bridging Into Customer Engagement

Once payments and hosting are sorted, the next challenge is keeping visitors engaged. That's where **chatbots** come in—your virtual assistants that answer questions, guide customers, and make your store feel alive.

In the next chapter, I'll walk you through choosing a chatbot platform that fits your business, whether you're starting small or planning to scale big.

Chapter 49: Choosing the Right Web Hosting for Your E-commerce Website

Web hosting is like the home for your website. It's where all your files, images, and pages live — and it's what keeps your store online for customers around the world. A reliable host ensures that your site loads quickly, stays secure, and doesn't crash when traffic spikes.

When you're running a store like *YourAIStudyBuddy.com* to sell *IntelliGloss: The Definitive AI Dictionary*, your hosting provider directly impacts how fast and smoothly customers can browse, learn, and buy.

When I launched my platform, I compared hosting options by looking at four main factors: **speed, pricing, security, and customer support**. Below are some of the most popular providers I explored, with notes on who they might be best for.

Popular Hosting Options for E-commerce Sites

- **Hostinger** — Affordable plans starting at $2.99/month, with free SSL certificates and WordPress optimization. I chose Hostinger for *YourAIStudyBuddy.com* because it gave me great loading times for global users (including Korea and the Netherlands). The only downside? The dashboard took some time to get used to.

- **SiteGround** — From $3.99/month. Offers managed WordPress hosting with strong security and excellent performance. A great option if you want peace of mind without too much manual setup.
- **Bluehost** — Beginner-friendly hosting from $2.95/month, with a free domain and SSL included. Their 24/7 support is especially helpful if you're not very technical.
- **WP Engine** — Premium hosting from $20/month. Known for lightning-fast speeds and high security. If you expect heavy traffic or want a "set it and forget it" option, this one's worth the higher cost.
- **DreamHost** — From $2.59/month, includes a free domain, unlimited bandwidth, and WordPress pre-installed. Very quick to launch if you're starting from scratch.
- **GoDaddy Hosting** — At $5.99/month, it works well if you're already using GoDaddy for your domain. Comes with WooCommerce and marketing tools built-in.
- **A2 Hosting** — From $2.99/month, famous for its "turbo servers" and speed optimization for WordPress. Perfect for stores where performance is a priority.
- **InMotion Hosting** — From $3.19/month, with a drag-and-drop site builder included. Beginner-friendly if you don't want to touch much code.
- **HostGator** — From $3.75/month, with unmetered bandwidth and 24/7 support. Reliable and affordable for small to medium-sized shops.
- **Cloudways** — Managed cloud hosting from $11/month. A flexible choice for scaling stores that may see sudden growth or seasonal spikes.
- **Kinsta** — From $30/month, built on Google Cloud infrastructure. Premium hosting for sites that demand top speed, reliability, and security.

Choosing the Best Host for *Your* Store

When picking a hosting provider, ask yourself:

1. **Where are my customers located?**
   The closer your hosting servers are to your audience, the faster your site will load for them.
2. **How much traffic do I expect?**
   If you're just starting out, a shared or budget plan works fine. But if you expect thousands of visitors quickly, a premium or managed host is worth it.
3. **How comfortable am I with tech?**
   If you want everything done for you, choose a managed WordPress host. If you don't mind tinkering, cheaper shared plans can work.

Final Thoughts

Your hosting provider is the foundation of your entire business. It determines how quickly your pages load, how safe your customer data is, and whether your site stays up during busy times.

I chose Hostinger because it balanced affordability with global performance. But what matters most is finding a host that matches *your* business needs, budget, and growth goals.

With the right hosting, you don't just put your store online—you give it the stability, speed, and security it needs to thrive.

Chapter 50: Website Builders — Flexible Alternatives to WordPress.org

When I built **YourAIStudyBuddy.com**, I chose **WordPress.org** because it gave me full control, flexibility, and a huge ecosystem of plugins like WooCommerce, Stripe, and Botpress. But I'll be honest—WordPress comes with a learning curve. For someone new to building websites, navigating themes, plugins, and hosting can feel overwhelming.

The good news? WordPress isn't your only option. There are many **website builders and CMS platforms** (Content Management Systems) that let you design and run a professional e-commerce store—even without coding. Some are visual, drag-and-drop tools; others are more advanced but offer powerful customization for those who want to grow big.

Here are some of the best alternatives, grouped by type, to help you choose the right fit for your project.

● Beginner-Friendly Platforms

These tools focus on **simplicity** and are great if you want to get online quickly without worrying about technical details.

- **Webflow** – A drag-and-drop platform that also functions as a CMS. You can design professional-looking sites visually, add products, and manage e-commerce without coding.
- **Joomla** – Open-source like WordPress but generally easier to manage. Comes with thousands of extensions for blogs, e-commerce, and even chatbots.
- **Concrete CMS** – Offers in-page editing, so you can make changes directly on your site and see updates instantly. Very beginner-friendly.

● Lightweight & Fast Platforms

These platforms are designed for speed and simplicity. They're great for small stores or projects where performance matters.

- **Grav** – A flat-file CMS (no database required), making it lightweight and fast. Ideal for small e-commerce sites or educational blogs.
- **ProcessWire** – Modular and flexible without being overwhelming. Lets you create clean product pages and structures without unnecessary clutter.

● Advanced / Scalable Platforms

These CMS options are better suited for **larger, more complex websites** with global audiences. They require more learning but give you more control.

- **Drupal** – Known for security and scalability. It's widely used by universities and global brands. Best for sites with complex user roles or large libraries of content.

- **Typo3** – Built for enterprise-level projects with multilingual support. Very powerful, but not ideal for beginners.
- **October CMS** – Runs on the Laravel framework, making it developer-friendly, fast, and extendable with plugins.
- **Craft CMS** – Offers a clean, modern interface for managing products and pages. Less cluttered than WordPress, but still powerful.
- **ExpressionEngine** – Modular and stable, with fewer plugins than WordPress, which can mean less maintenance.

My Experience

Even though I reviewed several of these platforms, I stayed with **WordPress.org** for YourAIStudyBuddy.com because:
✅ It had the widest range of e-commerce plugins
✅ It integrated seamlessly with Stripe for payments
✅ It worked well with Botpress for my AI chatbot
✅ The community support (tutorials, forums, YouTube videos) made problem-solving easier

But if I had wanted a **simpler, design-focused setup**, I might have chosen Webflow. If I wanted **enterprise-level scalability**, I might have leaned toward Drupal or Typo3.

Final Thoughts

The platform you choose should match your **goals, budget, and comfort level**. If you want fast setup and design flexibility, start with a visual builder like **Webflow**. If you want long-term scalability, explore **Drupal** or **Craft CMS**.

Remember: the platform is just the foundation. What matters most is how you use it to connect with your audience, organize your content, and bring your vision—just like I did with YourAIStudyBuddy.com—to life.

Chapter 51: Understanding Bots and Mastering Botpress for Your E-commerce Website

When I added my **AI Study Buddy chatbot** to *YourAIStudyBuddy.com*, it completely changed how I engaged with visitors. Suddenly, I had a tool that could:

- Answer questions about *IntelliGloss: The Definitive AI Dictionary*
- Guide users to my YouTube videos (like "What is a GPU?")
- Help students explore AI concepts in real time

As a non-technical entrepreneur who once struggled with WordPress menus and Hostinger's hidden settings, I found **Botpress** to be a game-changer. Its drag-and-drop design meant I could build an effective chatbot **without coding**.

This chapter will explain:

1. What a bot actually is.
2. Why chatbots matter for e-commerce.
3. How Botpress Studio helps beginners create and connect a chatbot to their site.

What Is a Bot?

Definition:
A *bot* (short for "robot") is a software program that automates tasks and interacts with users—often through conversation.

In practice:
On e-commerce sites, bots act like digital assistants. For example, on *YourAIStudyBuddy.com*, my bot can:

- Answer AI-related questions (like *"What is Machine Learning?"*)
- Direct users to products (such as IntelliGloss on Amazon)
- Provide support for account or order issues

Powered by AI and NLU:
Botpress uses **Natural Language Understanding (NLU)**, which means the bot can recognize different ways people ask the same question. Whether a visitor types "hi," "hello," or even a typo like "helo," the bot understands and responds.

Why Bots Matter for Your Website

Adding a chatbot isn't just "nice to have"—it directly impacts sales, engagement, and customer support:

- **Enhance User Experience** → Answer questions instantly, 24/7.
- **Boost Sales** → Guide users to buy IntelliGloss, upgrade subscriptions, or add extras.
- **Save Time** → Automate repetitive tasks (e.g., "Where can I download my worksheet?").
- **Global Reach** → Botpress supports 90+ languages, perfect for my audiences in Korea and the Netherlands.
- **Increase Engagement** → Encourage interaction, like prompting users to ask about AI terms, which also drives YouTube traffic.

Tips for Beginners Using Botpress

If you're new to chatbots, here's how I approached it:

1. **Start Simple** → Build a bot that answers the *top 3–5 common questions*. For me:
   - "Where can I buy IntelliGloss?"
   - "What's included in the Free Trial?"
   - "How do I contact support?"
2. **Leverage the No-Code Interface** → Use the drag-and-drop flow editor. It's like building conversation blocks with Lego—no coding needed.
3. **Test Often** → I started with a "HELLO" test. By asking my bot questions in different ways, I refined its answers and fixed gaps.

Key Takeaway

Think of your chatbot as your **first employee**—one that works tirelessly, greets every customer, and answers instantly. With Botpress, you don't need to be technical to build a bot that improves user experience, drives sales, and supports your community.

For *YourAIStudyBuddy.com*, my AI Study Buddy isn't just a chatbot—it's the link that connects books, videos, and students into a seamless learning experience.

Chapter 52: Choosing a Domain Name Company for Your E-commerce Website

When I launched **YourAIStudyBuddy.com** to sell *IntelliGloss: The Definitive AI Dictionary* and share short AI learning videos, I realized something important:
Your **domain name** is more than just a web address. It's your online identity, your brand, and the first impression visitors see. Choosing the right registrar (the company where you buy and manage your domain) is just as important as picking the name itself.

For me, I needed a domain that was:
✅ Short and memorable
✅ Easy to spell and share
✅ Globally trusted (.com)
✅ Supported by a registrar with good pricing, security, and integration

Here's what I learned while researching registrars—and the options you might consider.

🟢 Beginner-Friendly Registrars

These companies are easy to use, budget-friendly, and perfect if you're just starting out.

- **Hostinger** *(My Choice)*
  - Affordable domains (sometimes as low as $0.99/year) with free SSL on select plans.
  - Comes with hosting and WordPress integration in one dashboard.
  - Best for: Beginners who want hosting + domain in a single package.
- **Namecheap**
  - Transparent pricing, domains from $5.98/year, and free WHOIS privacy.
  - Clean, beginner-friendly interface.
  - Best for: Entrepreneurs who want low-cost domains without hidden fees.
- **Bluehost (through hosting bundles)**
  - Includes a free domain with hosting plans.
  - Strong support and WordPress focus.
  - Best for: Users who want one provider for everything.

🔵 Popular "One-Stop Shop" Registrars

These are great if you want domains plus extras like email, hosting, and marketing tools.

- **GoDaddy**
  - The world's largest registrar with 53+ million .coms registered.
  - Wide variety of extensions (.com, .shop, .ai, etc.).
  - Best for: Those who want domains, hosting, and marketing tools all in one place.
- **Name.com**
  - Professional extensions like .company, .shop, and .design.
  - Straightforward domain + DNS management.
  - Best for: Small businesses looking for polished, professional branding.

● Security-Focused Registrars

If protecting your brand and customers is your top priority, these options shine.

- **Cloudflare Registrar**
  - Sells domains at wholesale cost—no markup.
  - Built-in security like free SSL, DDoS protection, and DNS management.
  - Best for: Security-conscious businesses or global stores.
- **Google Domains (now part of Squarespace)**
  - Simple, reliable, and integrates with Google Workspace for email.
  - Domains start at $7/year.
  - Best for: Users already using Gmail, Google Drive, or Google Workspace.
- **DNSimple**
  - Clean interface, efficient DNS management, and great automation.
  - Plans from $8/year.
  - Best for: Businesses managing multiple domains and subdomains (like kr.youraistudybuddy.com for Korea).

○ Creative & Global-Friendly Registrars

These are ideal if you want to stand out with unique or country-specific extensions.

- **Dynadot**
  - Offers 500+ extensions (like .ai, .io, .tech, and country-specific domains).
  - Includes a marketplace for buying and selling premium domains.
  - Best for: AI-focused or international brands.
- **In-country registrars**
  - If you're targeting a specific region (e.g., .kr for Korea, .nl for the Netherlands), sometimes buying directly through a local registrar builds trust and improves SEO.
  - Best for: Businesses expanding into specific countries.

Why Domain Choice Matters for E-commerce

- **Trust:** A .com domain is still the gold standard—99% of Fortune 500 companies use it.
- **Clarity:** A clear, simple name helps customers remember and return.
- **Local reach:** Country-code domains (.kr, .nl, .uk) can increase trust with local customers.
- **Branding:** Niche TLDs like .ai or .tech highlight your focus and make your brand more memorable.

For YourAIStudyBuddy.com, I chose **.com** for global credibility. But I'm also considering **.ai** for branding and **.kr/.nl** for local audiences, since my Botpress chatbot supports multiple languages.

Beginner Tips

1. Always register the **.com** if it's available.
2. Add country-specific domains if you want to localize your store.
3. Use free WHOIS privacy to protect your personal info.
4. Keep your domain and hosting **with the same provider** if you want simplicity—or separate them if you want more control.

✨ **Final Thought:**
Your domain name is your brand's front door. Choose a registrar that offers security, simplicity, and room to grow. Whether you start with Hostinger for budget-friendliness, Cloudflare for security, or Dynadot for creative TLDs, the right choice will make your e-commerce journey smoother and your brand stronger.

Chapter 53: Hostinger Dashboard Guide for Building E-commerce Websites

When I built **youraistudybuddy.com** to sell *IntelliGloss: The Definitive AI Dictionary* and share my two-minute YouTube videos on AI terms like "What is a GPU?", I found Hostinger's interface both powerful and frustrating. As a non-techy entrepreneur, I struggled with unclear labels and hidden settings—much like my early challenges with WordPress.

This chapter breaks down Hostinger's **hPanel dashboard** (the control center for your website) into clear, beginner-friendly sections. Whether you're launching an online store for books like *IntelliGloss* or adding a chatbot like my AI Study Buddy, these explanations will help you master Hostinger and build a professional site that shines globally—just as mine now reaches customers in over 20 countries.

First Things First: Domains and Hosting

Before we dive into Hostinger's dashboard, let's clear up two essentials:

What Is a Domain Name?
A domain name is your website's address on the internet. It's what people type into their browser to visit your site—for example, *youraistudybuddy.com*. Just like your home has a street address, your website needs a domain so people can find it.

Examples of domain names include:

1. google.com
2. amazon.com

3. halasflowers.com (if I built a website just for my flower shop)

Every domain name is unique, and you usually purchase it from a registrar like GoDaddy, Namecheap, or Hostinger (many hosting companies sell domains directly).

What Is a Hosting Company?
Think of hosting as the land and house where your website "lives." Even if you own a domain name (the address), your site needs a place to store its pages, images, videos, and files. That's what a hosting company provides.

The hosting company:

1. Stores your website's files on a server
2. Keeps your site online 24/7
3. Provides tools to manage your website (like installing WordPress, adding SSL, and creating email addresses such as info@youraistudybuddy.com)

In short:

- **Domain name = street address**
- **Hosting company = land and house**
- **WordPress = interior designer** that helps you build the inside.

In my case, I chose **Hostinger** because it included free email, built-in security, and one-click WordPress installation.

Navigating the Hostinger Dashboard

Here's a guided tour of the main sections in Hostinger's **hPanel** and why they matter for e-commerce.

1. Home

What It Does
The Home (Dashboard) section is your starting point after logging in. It shows an overview of your websites, hosting plans, domains, and recent activity—such as new orders or performance alerts.

Why It's Important
Think of it as your store's front desk. It lets you monitor site uptime, track updates, and access shortcuts like "Add Website," "Manage Domains," or "Check Emails."

Tips for Beginners

- Use **Add Website** to launch a WordPress store with WooCommerce.
- Check the **To-Do List** for setup tasks (like enabling SSL for secure payments).
- Review the **Performance widget** to keep your site fast worldwide.

2. Websites

What It Does
This lists all sites under your hosting plan. From here, you can install WordPress, manage plugins, set up WooCommerce, or add tools like Botpress for chatbots.

Why It's Important
It's the "control room" for building and managing your store. I used it to install WordPress, add WooCommerce, and integrate my chatbot.

Tips for Beginners

- Use the **Auto Installer** for plugins like Yoast SEO.
- Think of this as the hub where every key step begins.

3. Domains

What It Does
Manage, register, or connect domains. You can also set up subdomains (e.g., blog.youraistudybuddy.com).

Why It's Important
A domain is your brand's digital address. This is where you secure and connect it to your store.

Tips for Beginners

- Pick a short, brand-relevant domain.
- Update DNS carefully—mistakes can make your site disappear.
- Use subdomains for specialized sections like blogs or shops.

4. Hosting

What It Does
Shows plan details (e.g., Business or Cloud Startup) and lets you manage resources like storage, bandwidth, and backups.

Why It's Important
Reliable hosting ensures your store runs smoothly, even during traffic spikes.

Tips for Beginners

- Choose a WooCommerce-optimized plan (often under $5/month).
- Enable LiteSpeed Cache for faster load times.
- Monitor storage to ensure you can add product images and videos.

5. Emails

What It Does
Create professional addresses like sales@youraistudybuddy.com and manage email forwarding or webmail.

Why It's Important
Professional email builds customer trust and supports e-commerce (e.g., order confirmations).

Tips for Beginners

- Create up to 100 free emails with Hostinger plans.
- Configure WP Mail SMTP so emails don't land in spam.
- Check webmail regularly for inquiries.

5. Online Store

What It Does
Guides you through setting up WooCommerce: store name, product types, payments, and shipping.

Why It's Important
This section simplifies launching your store, especially for beginners.

Tips for Beginners

- Follow the checklist (name, product type, payment gateways).
- Add products like physical books or digital downloads.
- Use analytics to track sales and improve listings.

7. SEO

What It Does
Provides tools for search engine optimization—meta titles, descriptions, keywords, and sitemap generation.

Why It's Important
SEO brings customers to your store. Optimized titles helped my *IntelliGloss* rank higher on Google.

Tips for Beginners

- Keep titles under 60 characters, descriptions under 160.
- Use keywords like "AI dictionary" or "learn GPU."
- Submit your sitemap to Google Search Console.

8. Security

What It Does
Manages SSL, 2FA, malware scanning, and DDoS protection.

Why It's Important
Security protects customer data and builds trust.

Tips for Beginners

- Enable the free SSL certificate.
- Use two-factor authentication.
- Run malware scans weekly.

9. File Manager

What It Does
A web tool for managing site files. You can upload images, edit configuration files, or back up without FTP.

Why It's Important
Helps organize files and troubleshoot issues.

Tips for Beginners

- Use wp-content/uploads for product images.
- Always back up before editing.
- If it feels overwhelming, install File Manager Pro in WordPress.

10. Databases

What It Does
Lets you create and manage MySQL databases. Access phpMyAdmin if you need direct edits.

Why It's Important
Databases store your products, orders, and customer info.

Tips for Beginners

- Let Hostinger auto-create databases during WordPress install.
- Avoid editing manually unless guided by support.
- Back up databases regularly.

11. Billing

What It Does
Tracks your hosting payments, renewals, and invoices.

Why It's Important
Keeping billing current prevents downtime.

Tips for Beginners

- Choose plans that include free domains and SSL.
- Set reminders for renewal dates.
- Contact support if billing terms are unclear.

Bonus: Additional Tools After Setup

Hostinger Tools Plugin

Appears in your WordPress dashboard. Includes Onboarding checklists, cache management, and an AI Content Creator.

Why It's Important
I used it to configure WooCommerce and write SEO-friendly descriptions faster.

Tips for Beginners

- Follow the Onboarding steps to complete setup.
- Use AI Content Creator to generate blog posts tied to your YouTube videos.

WooCommerce

Appears in WordPress after installation. Lets you manage products, orders, and analytics.

Why It's Important
This is your e-commerce engine. I used it to sell *IntelliGloss*, set up Stripe payments, and track sales globally.

Tips for Beginners

- Complete the WooCommerce wizard for quick setup.
- Use Analytics to study sales trends and optimize.

Wrap-Up: Your Next Step

Before building your dream store in WordPress, secure your **domain name** and choose a **hosting company**. They are the foundation—your address and your home. From there, Hostinger's dashboard becomes your command center for running a global e-commerce business.

In the next chapter, I'll walk you through how to connect your domain name to your hosting and install WordPress—the step where your website finally comes alive.

Chapter 54: Understanding WordPress.org for Non-Tech Website Builders

When I built **youraistudybuddy.com** to sell *IntelliGloss: The Definitive AI Dictionary* and integrate my AI Study Buddy chatbot, I discovered the power of **WordPress.org.** It's a free, open-source platform that gives entrepreneurs like us the ability to build professional, scalable websites without hiring developers.

At first, WordPress's dashboard felt overwhelming—menus, settings, and hidden options left me frustrated. But once I understood how the system works, it became my most valuable tool. This chapter will help you master WordPress.org, avoid common pitfalls, and build with confidence—even if you don't have a tech background.

What Is WordPress.org?

What It Is
WordPress.org is a free content management system (CMS) powering over 40% of the world's websites—from small blogs to global e-commerce stores. Unlike **WordPress.com** (which limits customization and hosts your site for you), WordPress.org gives you **full control.** You host it yourself (on platforms like Hostinger), and you can install themes, plugins, and even custom code if you choose.

Why It Matters
For non-technical builders, this flexibility is a game-changer. You can design professional sites, sell products, add subscriptions, and integrate AI tools like chatbots—all without coding.

What You Must Understand
The WordPress ecosystem is massive:

- **WooCommerce** for e-commerce
- **WP Simple Pay** for subscriptions
- **Botpress** for chatbots
- **Yoast SEO** for search optimization

With the right combination, you can run a global business like I do, reaching 20+ countries.

Navigating the WordPress Dashboard

What It Is
The WordPress dashboard is your **control center.** The left-hand menu organizes your site's core functions:

- **Posts** – for blogs and updates
- **Pages** – for static sections like Home or Terms & Conditions
- **Appearance** – for themes, menus, and widgets
- **Plugins** – for adding features

- **Users** – for account management
- **Tools** – for backups and imports
- **Settings** – for site-wide configuration

Why It Matters
If you understand this layout, you won't get lost or misconfigure key features.

What You Must Understand
Use the **Screen Options** tab (top right) to hide clutter. Explore each menu slowly. For example:

- I built my IntelliGloss homepage under **Pages**
- Added subscription tools with **WP Simple Pay** under **Plugins**
- Linked Stripe payments through plugin settings

What Can Go Wrong
Rushing through the dashboard can lead to broken links, confusing menus, or plugins that don't work well together.

Setting Your Homepage Correctly

What It Is
Your homepage is your digital storefront.

Why It Matters
If misconfigured, visitors may land on the wrong page—or worse, get stuck in a redirect loop (as happened to me).

What You Must Understand

- Go to **Settings → Reading**
- Select *"A static page"* and choose the correct Home page
- Clear cache in Hostinger's hPanel if issues persist
- Test in an incognito browser window

What Can Go Wrong
Plugin conflicts (like WP Simple Pay or Botpress) can override homepage settings. Always test after changes.

Connecting Pages Seamlessly

What It Is
Every page on your site—Home, Products, Plans, Login—must be properly linked.

Why It Matters
Broken or misdirected links frustrate visitors and cost you sales.

What You Must Understand

- Use **Appearance → Menus** to connect pages

- Copy correct URLs (e.g., youraistudybuddy.com/starter-plan-payment)
- For subscriptions, embed payment forms using shortcodes (e.g., `[simpay id="89"]`)

What Can Go Wrong
A wrong link or outdated URL can make your "Buy Now" button lead nowhere, as I experienced. Always test your navigation as a user would.

Choosing Themes and Plugins

What It Is
Themes control design. Plugins add functionality.

Why It Matters
The right combination creates a fast, professional site.

What You Must Understand

- Use lightweight themes like **Astra** or **Storefront**
- Essential plugins include:
    - **WooCommerce** (store)
    - **WP Simple Pay** (subscriptions)
    - **Botpress** (AI chatbot)
    - **Yoast SEO** (search optimization)
- Limit plugins to what you need (quality > quantity).

What Can Go Wrong
Too many or outdated plugins can slow your site or cause conflicts.

Setting Up E-Commerce and Subscriptions

What It Is
WordPress integrates seamlessly with WooCommerce and Stripe-based tools for e-commerce.

Why It Matters
This is how you sell products, offer plans, and build recurring revenue.

What You Must Understand

- Install **WooCommerce** and run the setup wizard
- Use **WP Simple Pay** for subscription plans
- Test in Stripe's sandbox mode before going live

What Can Go Wrong
Missed steps in setup can break checkout flows. Always test with sample transactions first.

Keeping WordPress Updated and Secure

What It Is
WordPress regularly releases updates for its core, themes, and plugins.

Why It Matters
Updates improve speed, fix bugs, and patch security holes.

What You Must Understand

- Check **Dashboard → Updates** weekly
- Back up your site before updating (via Hostinger's backup tool)
- Use a security plugin like **Wordfence**

What Can Go Wrong
Skipping updates leaves your site vulnerable to hacks or crashes—especially dangerous for payment pages.

Wrap-Up
WordPress.org may feel intimidating at first, but once you master its dashboard, it becomes your most powerful ally. Combined with the hosting foundation you'll explore in the next chapter, it allows you to create an AI-powered e-commerce site that's both professional and global.

Chapter 55: The 7 Core Players Behind Every AI-Powered E-Commerce Website

Every successful AI-powered e-commerce website rests on a set of essential players working together behind the scenes. Without one of them, your site either won't launch, won't function properly, or won't grow. Let's break down the seven core players and why they matter.

1. Domain Registrar

What It Is
A domain registrar is the company that sells you your website's address—your domain name (e.g., *youraistudybuddy.com*). Think of it as buying the deed to a piece of digital land.

Why It Matters
Without a domain, your site has no official name or URL on the internet. It's the starting point of your online identity.

What You Must Understand
Registrars like GoDaddy, Namecheap, and Google Domains also handle DNS settings, renewals, and privacy protection. Beginners can register a name in just minutes.

What Can Go Wrong
If your domain isn't renewed on time, it can expire, leaving your site offline—or worse, someone else could buy it.

2. Web Hosting Provider

What It Is
A hosting provider stores your website's content—everything from text and images to plugins and databases—and serves it to visitors when they access your site.

Why It Matters
No hosting = no website. Hosting ensures your site is secure, fast, and always available.

What You Must Understand
Think of your website like a house: the domain name is the address, and hosting is the land it's built on. Providers like Hostinger, Bluehost, and SiteGround offer beginner-friendly options that integrate with WordPress.

What Can Go Wrong
Cheap or unreliable hosting can lead to downtime, slow speeds, and poor customer experiences—especially during peak sales periods.

3. Content Management System (CMS)

What It Is
A CMS is a tool for building, managing, and organizing your website without coding. WordPress.org is the most popular and powerful option, and it's free.

Why It Matters
It's your control center—the canvas where you design pages, upload content, and manage users.

What You Must Understand
WordPress is flexible and scalable. You can build anything from a simple blog to a global online store.

What Can Go Wrong
Failing to update WordPress regularly can leave your site vulnerable to hacks and performance issues.

4. E-Commerce Plugin or Platform

What It Is
An e-commerce platform adds the ability to sell products, process orders, and manage a store. For WordPress, that platform is usually WooCommerce.

Why It Matters
Without an e-commerce engine, you can't list products, accept payments, or run your business.

What You Must Understand
WooCommerce integrates seamlessly with WordPress, while hosted solutions like Shopify or BigCommerce bundle everything together.

What Can Go Wrong
Poor configuration can result in failed checkouts, incorrect tax settings, or missing product pages.

5. AI Bot Platform

What It Is
An AI bot platform like **Botpress** provides automated, intelligent customer support. It's like having a 24/7 virtual assistant built into your store.

Why It Matters
AI bots save time, reduce support costs, and improve customer satisfaction by answering FAQs, handling orders, or suggesting products.

What You Must Understand
Platforms like Botpress allow visual flow building and integrations with Stripe, WooCommerce, or email tools. Bots can answer "Where's my order?" or "What's your return policy?" while also learning from interactions to improve responses.

What Can Go Wrong
Poorly trained bots frustrate customers and hurt trust. Always test your bot before going live.

6. Payment Gateway

What It Is
A payment gateway securely processes credit cards and online payments. Stripe, PayPal, and Authorize.Net are popular options.

Why It Matters
Without it, you can't accept money online—making it the heartbeat of your sales system.

What You Must Understand
Gateways encrypt data, verify transactions, and send funds to your bank. Stripe also supports subscriptions, invoices, and refunds.

What Can Go Wrong
Incorrect setup can cause failed payments, chargebacks, or even account suspension.

7. Email Marketing / SMTP Service

What It Is
Email tools ensure your system and marketing emails reach customers. WP Mail SMTP improves deliverability, while platforms like Mailchimp, Brevo, or ConvertKit help with newsletters and campaigns.

Why It Matters
Email builds trust, loyalty, and conversions. From order confirmations to product launches, it's your direct line to customers.

What You Must Understand
WordPress alone is unreliable at sending emails. You need SMTP authentication and marketing platforms to keep communication professional.

What Can Go Wrong
If not configured correctly, emails land in spam—or never arrive at all.

Wrap-Up
These seven players form the backbone of every AI-powered e-commerce site. The domain and hosting give you an address and a home. The CMS and e-commerce engine let you build and sell. AI bots enhance customer experience. Payment gateways ensure money flows securely. And email services keep you connected. With these in place, you're ready to build the actual **pages and content** that bring your online store to life.

Chapter56 : Recap – Building the Foundation of Your Website

Once you've secured your domain, hosting, and core tools, it's time to build the visible structure of your site. These steps create the foundation your customers will interact with daily.

1. Create Pages

What It Is
Pages are permanent parts of your site, such as your Homepage, Login Page, Products Page, or Terms & Conditions.

Why It Matters
Pages define your website's structure. Without them, visitors get confused or may even doubt your credibility.

What You Must Understand
Every page plays a role: the Home page introduces your brand, the Products page generates revenue, and legal pages keep you compliant. In WordPress, pages are managed under the "Pages" tab.

What Can Go Wrong
Missing or poorly designed pages lead to "404 Not Found" errors or rejected payment integrations (e.g., Stripe may require Terms & Conditions).

2. Write Blog Posts

What It Is
Blog posts are time-sensitive entries (tutorials, updates, news) that appear in your blog feed.

Why It Matters
Fresh content drives traffic from search engines and builds authority.

What You Must Understand
Posts are organized by date, category, and tags. They're managed under "Posts" in WordPress.

What Can Go Wrong
Inactive or poorly written blogs make your site look abandoned and harm SEO.

3. Upload Images

What It Is
Images are added through the Media Library and can be placed in pages, posts, or products.

Why It Matters
Visuals capture attention, explain ideas faster, and increase engagement.

What You Must Understand
Rename files with keywords (e.g., *ai-dictionary-cover.jpg*) and add alt-text for SEO. Compress large images for faster loading.

What Can Go Wrong
Unoptimized images slow down your site and hurt rankings.

4. Approve or Delete Comments

What It Is
Comment management allows you to approve, trash, or mark posts as spam.

Why It Matters
Comments build community and credibility—but also attract spam.

What You Must Understand
WordPress has built-in moderation tools, and plugins like Akismet filter spam automatically.

What Can Go Wrong
Leaving spam visible damages trust and can lower search rankings.

5. Install Plugins and Themes

What It Is
Plugins add features (e.g., SEO, security, payments). Themes define your site's design and layout.

Why It Matters
Plugins give functionality; themes set the look and feel. Both are essential.

What You Must Understand
Search, install, and update plugins carefully. Test themes in preview before activating.

What Can Go Wrong
Bad or outdated plugins/themes can crash your site or create security vulnerabilities.

6. Connect Email Tools

What It Is
Tools like WP Mail SMTP ensure reliable email delivery, while platforms like Mailchimp handle newsletters.

Why It Matters
Without this, customers may never receive receipts, password resets, or updates.

What You Must Understand
Configure DNS records (SPF, DKIM, DMARC) for authentication and test before launch.

What Can Go Wrong
Emails could land in spam or fail to deliver entirely, hurting trust.

7. Add Payment Systems

What It Is
Integrations like Stripe or PayPal allow customers to pay securely.

Why It Matters
Payments are the lifeline of your business.

What You Must Understand
Link your bank account, enable live/test modes, and comply with tax/refund laws.

What Can Go Wrong
Improper setup leads to failed checkouts or even account suspension.

8. Set Up Menus

What It Is
Menus are your site's navigation links (Home, Blog, Shop, Contact, etc.).

Why It Matters
Menus guide visitors. A confusing layout makes users leave.

What You Must Understand
Menus are edited under "Appearance → Menus" in WordPress. You can create dropdowns, headers, or footer menus.

What Can Go Wrong
Broken or missing links frustrate users and drive them away.

9. Secure Your Site

What It Is
Site security includes firewalls, backups, strong passwords, and anti-malware.

Why It Matters
A secure site protects your customers, data, and reputation.

What You Must Understand
Use plugins like Wordfence, enable 2FA, and schedule automatic backups.

What Can Go Wrong
Hacks, data loss, or downtime can destroy trust and revenue.

10. Connect to Stripe

What It Is
Stripe is a payment processor that lets you accept credit cards directly on your site.

Why It Matters
It's one of the most trusted solutions for online businesses.

What You Must Understand
Integrate Stripe through plugins or APIs, test in Sandbox mode, and activate Live mode once ready.

What Can Go Wrong
Improper integration blocks sales, causes errors at checkout, or prevents accurate reporting.

Wrap-Up
By completing these ten steps, your site becomes more than just an address and a host—it becomes a living, functioning store. From pages and posts to payments and security, every element matters. Together, they create a seamless experience for your visitors and a solid foundation for your e-commerce success.

Chapter 57: Understanding Hostinger for Non-Tech Website Builders

When I chose **Hostinger** to host **youraistudybuddy.com**, I was drawn by its affordability and simplicity. As a non-technical entrepreneur, I needed a platform that made website building possible without coding. Hostinger delivered—though I faced navigation quirks along the way. This chapter will show you how Hostinger works, why it's beginner-friendly, and how to use it to build and scale your own online store.

Who Is Hostinger?

What It Is
Hostinger is a global hosting company founded in 2004 in Lithuania, now serving millions of users in 150+ countries.

Why It Matters
Its focus is on affordability, simplicity, and user-friendly design—perfect for beginners, educators, and entrepreneurs.

What You Must Understand
Hostinger is designed for non-tech builders, with:

- Clean dashboards (hPanel)
- 24/7 customer support

- Built-in AI tools
- Affordable plans that grow with your business

Navigating hPanel

What It Is
Hostinger's control panel (**hPanel**) is your hosting command center.

Why It Matters
It simplifies hosting tasks that would otherwise require coding.

What You Must Understand
Key sections include:

- **Websites** – add/manage WordPress or Website Builder sites
- **Domains** – register or connect domains
- **Hosting** – view plan details and upgrades
- **Emails** – create professional addresses
- **Databases** – manage WordPress data
- **Security** – SSL, malware scans

What Can Go Wrong
Navigation can feel confusing at first—use the search bar or live chat support when stuck.

Setting Up a WordPress Site with Hostinger

What It Is
Hostinger offers one-click WordPress installation with LiteSpeed servers for speed.

Why It Matters
It's the easiest way for beginners to launch a WordPress site.

What You Must Understand

- Go to **hPanel → Websites → Add Website**
- Select WordPress and follow the wizard
- Access your site at *domain.com/wp-admin*

What Can Go Wrong
If redirects occur, clear cache (**hPanel → Cache → Purge All**) and check WordPress homepage settings (see Chapter 45).

Using Hostinger's Website Builder

What It Is
A drag-and-drop, no-code builder powered by AI.

Why It Matters
It's perfect for those who want a site quickly without WordPress complexity.

What You Must Understand

- Go to **hPanel → Websites → Add Website → Website Builder**
- AI generates templates and text in minutes
- Add e-commerce features under "Online Store"

What Can Go Wrong
The Website Builder doesn't support nested URLs (bad for SEO). For global e-commerce, WordPress is usually better.

Configuring E-Commerce and Subscriptions

What It Is
Hostinger supports e-commerce through both WordPress and its Website Builder.

Why It Matters
This is how you turn your site into a store.

What You Must Understand

- Use **WooCommerce** or **WP Simple Pay** for subscriptions
- Connect to Stripe for secure payments
- Test in Stripe's sandbox mode before launch

What Can Go Wrong
Incorrect shortcode use or Stripe settings may break your checkout page.

Leveraging Hostinger's AI Tools

What It Is
Hostinger offers AI-powered helpers like AI Writer, AI SEO Assistant, and AI Heatmap.

Why It Matters
AI tools save time and improve performance for non-technical users.

What You Must Understand

- Use **AI Writer** for quick blog posts or product descriptions
- Use **SEO Assistant** for meta tags and rankings
- Use **Heatmap** to track visitor clicks and improve page layouts

What Can Go Wrong
AI-generated content may feel generic. Always edit for personal tone.

Keeping Hostinger Secure and Updated

What It Is
Hostinger provides updates, SSL, and malware scans.

Why It Matters
Security is critical for e-commerce sites handling payments.

What You Must Understand

- Enable free SSL (adds "https://")
- Use **hPanel → Backups** regularly
- Check for updates weekly

What Can Go Wrong
Ignoring updates risks downtime or vulnerabilities.

Integrating AI Chatbots

What It Is
You can embed chatbots like **Botpress** into Hostinger-powered sites.

Why It Matters
AI bots engage customers, answer questions, and guide them to sales pages.

What You Must Understand

- In WordPress: install Botpress plugin
- In Website Builder: use **Integrations → Custom Code**

What Can Go Wrong
Poorly trained bots frustrate visitors. Test before going live.

Benefits of Using Hostinger

- Beginner-friendly dashboards
- Affordable plans (often under $5/month)
- Excellent uptime and speed
- AI tools and SEO support
- Scales easily as your site grows

Wrap-Up
Hostinger is more than just a hosting company—it's a launchpad for non-tech entrepreneurs. Combined with WordPress (Chapter 45), it provides everything you need to build, run, and scale an AI-powered e-commerce business. My journey proves you don't need a coding background—just the right tools, patience, and persistence.

Chapter 58: Understanding Botpress for Non-Tech Website Builders

Understanding Botpress for Non-Tech Website Builders

Botpress powers the AI Study Buddy chatbot on my e-commerce site, youraistudybuddy.com, guiding customers to buy *IntelliGloss: The Definitive AI Dictionary* across 20+ countries and directing them to subscription pages like my $9.95/month Starter Plan. As a non-technical entrepreneur, I found Botpress's no-code platform a game-changer, though setting up flows for subscriptions took trial and error due to navigation quirks. This chapter explains who Botpress is, why a 24/7 chatbot is vital for e-commerce, and how to add one to your site without coding. With Botpress, you can automate customer support, boost sales, and create a seamless experience, even if you've never built a chatbot before, just as I did for my global audience.

Who Is Botpress and What It Does

Botpress is an open-source, no-code platform for building AI-powered chatbots, founded in 2017 in Quebec, Canada, to make conversational AI accessible to businesses. It's used by companies like Windstream to handle customer queries, automate tasks, and integrate with platforms like WordPress, Shopify, and Stripe. For youraistudybuddy.com, my AI Study Buddy chatbot answers FAQs (e.g., "What's IntelliGloss?"), suggests plans, and links to payment pages like `youraistudybuddy.com/starter-plan-payment` with `[simpay id="89"]`. Botpress's drag-and-drop studio and natural language processing (NLP) let non-technical users create human-like chatbots that work 24/7, never call in sick, and scale to millions of interactions, perfect for small businesses or educators selling globally. *See Glossary: AI Chatbot, NLP.*

Why a Chatbot Is Essential for Your E-commerce Site

A chatbot like Botpress is a virtual assistant that transforms e-commerce by providing instant, 24/7 customer support, unlike human agents who need breaks. For youraistudybuddy.com, my AI Study Buddy handles inquiries across time zones, ensuring customers in 20+ countries get help anytime, boosting satisfaction by 3%, as seen in cases like Photobucket. Chatbots save costs—reducing support tickets by up to 70%—and drive sales by guiding users to checkouts, like my Family Basic plan. They use NLP to understand natural language, making interactions feel personal, not robotic. In a competitive market, a chatbot's constant availability and personalization (e.g., recommending *IntelliGloss* based on browsing) keep customers engaged, increasing conversions and loyalty. *See Glossary: Customer Satisfaction (CSAT).*

Benefits of Using Botpress for Your Site

Botpress offers powerful benefits for non-technical builders. Its free, open-source plan includes core features like a drag-and-drop flow editor, supporting unlimited users on youraistudybuddy.com. Paid plans (starting at $89/month) add analytics and enterprise integrations. It handles multiple chats simultaneously, scaling for peak traffic, unlike human agents. For *IntelliGloss*, my chatbot suggests plans and recovers abandoned carts, boosting sales. Botpress integrates with WordPress, Stripe, and messaging apps (e.g., WhatsApp), enabling seamless payments and global reach in 90+ languages. Its analytics track user intent, helping refine responses, and GDPR-compliant security protects customer data, critical for my subscription pages. Plus, it never calls in sick, ensuring 24/7 reliability for your store.

Setting Up a Botpress Chatbot for WordPress

Setting up Botpress for youraistudybuddy.com is simple, even for non-coders. Sign up at `botpress.com`, choose the free plan, and create a chatbot in **Botpress Studio → Create New Bot**. Use the drag-and-drop editor to build flows (e.g., "Greet user" → "Ask about IntelliGloss" → "Link to Starter Plan"). Train the bot with your site's content (**Knowledge Base → Add Website URL**: `youraistudybuddy.com`) to answer FAQs. In WordPress, install the Botpress plugin (**Plugins → Add New → Botpress**), copy the bot's embed code from **Botpress Studio → Integrations → Webchat**, and paste it into your site's header (**Appearance → Theme File Editor → header.php**). Test the chatbot on pages like `youraistudybuddy.com/starter-plan-payment` to ensure it links to `[simpay id="89"]`. I struggled with flow navigation, so use Botpress Academy tutorials for guidance. *See Glossary: Knowledge Base.*

Integrating Botpress with Stripe for Subscriptions

Botpress enhances subscriptions by guiding users to payment pages, as I did for my $9.95/month Starter Plan. In **Botpress Studio → Flows**, create a flow: "Ask about plans" → "Suggest Starter Plan" → "Send URL: `youraistudybuddy.com/starter-plan-payment`". Connect Botpress to Stripe via webhooks (**Botpress Studio → Integrations → Webhooks**) to trigger actions (e.g., confirm payment success). In WordPress, ensure WP Simple Pay forms (e.g., `[simpay id="90"]` for Family Basic) are linked correctly. Test in Stripe's Test Mode (card: `4242 4242 4242 4242`) to verify the chatbot directs users to the right page without redirect errors, which I fixed by clearing Hostinger's cache (**hPanel → Cache → Purge All**). This setup drives *IntelliGloss* sales seamlessly. *See Glossary: Webhook, Subscription Plan.*

Enhancing Engagement with Botpress's AI Features

Botpress's NLP makes my AI Study Buddy feel like a real assistant, answering complex queries like "What's the difference between Starter and Family plans?" It uses customer data to personalize responses, recommending *IntelliGloss* or YouTube tutorials based on browsing history. In **Botpress Studio → Intents**, train intents (e.g., "plan_details" → "Describe Starter Plan features") to handle FAQs. For global reach, enable multilingual support (**Settings → Languages**) for 20+ country audiences. Add proactive triggers (**Flows → Proactive Messages**) to greet users on your "Choose Your Plan" page, like "Need help picking a plan?" This increased my conversions by engaging users early, as seen in e-commerce cases where chatbots recover 70% of abandoned carts. *See Glossary: AI Agent.*

Troubleshooting Common Botpress Issues

Botpress setup can have hiccups, like my initial misdirected chatbot links. If the bot doesn't load, check the embed code in WordPress (**Appearance → Theme File Editor**) and ensure the Botpress plugin is active. If responses are off, retrain the knowledge base (**Botpress Studio → Knowledge Base → Retrain**) with updated site content. For subscription page errors, verify URLs in flows and test redirects in an incognito window. If integration with Stripe fails, recheck webhook settings (**Botpress Studio → Integrations**). Botpress's Discord community (25,000+ members) and YouTube tutorials helped me fix flow errors quickly. Contact support via `botpress.com/support` for complex issues, ensuring your *IntelliGloss* chatbot runs smoothly. *See Glossary: Redirect.*

Optimizing Botpress for Global E-commerce

Botpress shines for global stores like youraistudybuddy.com, supporting 90+ languages to engage customers in their native tongue. In **Botpress Studio → Settings → Languages**, enable auto-detection for users in Europe or Asia. Integrate with Stripe to handle 135+ currencies, syncing with your subscription plans. Use analytics (**Botpress Studio → Analytics**) to track metrics like conversion rates and user drop-offs, refining flows to reduce my 70% cart abandonment rate. For example, I added a "Low stock alert" message for *IntelliGloss*, prompting faster checkouts. Botpress's geolocation feature can offer location-based deals, like discounts for

users near educational hubs, enhancing your global reach and loyalty. *See Glossary: Conversion Rate.*

Keeping Your Botpress Chatbot Updated and Secure

Botpress's open-source nature ensures frequent updates, but you must maintain them. Check for updates in **Botpress Studio → Settings → Updates** monthly to access new features like enhanced NLP. Back up flows (**Botpress Studio → Export**) before updates to avoid data loss, a lesson I learned when a flow broke my AI Study Buddy. Enable GDPR-compliant settings (**Settings → Compliance**) to protect customer data, vital for my subscription pages. Use **Analytics → Conversation Insights** to monitor user interactions and fix outdated responses. Botpress's security suite and 24/7 community support keep your chatbot reliable, ensuring *IntelliGloss* customers trust your site.

Tips for Non-Technical Success with Botpress

Botpress empowered me to build a 24/7 AI Study Buddy without coding, but its flow editor took practice to master. Start with simple flows (e.g., greet → FAQ → payment link) and test in **Botpress Studio → Emulator**. Use Botpress Academy or WPBeginner for tutorials, and limit flows to 10–15 to avoid complexity. Test your chatbot as a customer in an incognito window to catch errors, like my mislinked payment page. For *IntelliGloss*, I focused on one goal—guiding users to subscriptions—before adding features like multilingual support. Your chatbot doesn't need to be perfect; prioritize engagement, then refine to match your brand's voice, whether selling books or courses globally.

Why Botpress Is Your Path to E-commerce Success

Botpress's no-code platform, 24/7 availability, and global capabilities make it ideal for non-technical builders like me, who created a dynamic chatbot for youraistudybuddy.com. It never calls in sick, handles unlimited queries, and boosts sales by guiding users to *IntelliGloss* subscriptions. By mastering its drag-and-drop studio, integrating with Stripe, and leveraging AI features, you can build a professional chatbot that grows your brand. My journey—from AT&T and flower shops to an AI education leader—proves Botpress's tools, despite setup learning curves, empower anyone to succeed. Take it step by step, test thoroughly, and soon you'll have a chatbot that connects with customers worldwide, just like my AI Study Buddy does.

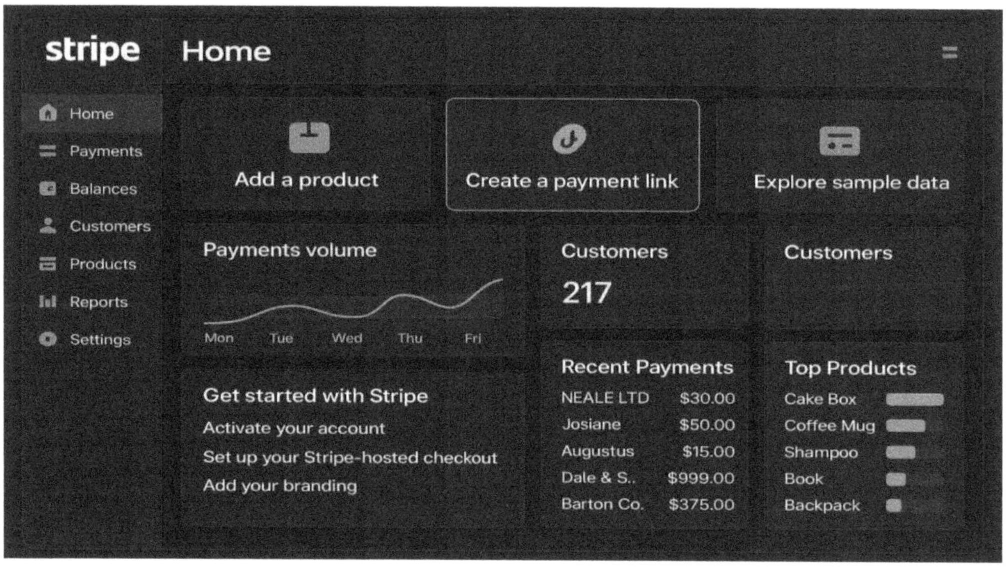

Chapter 59: Understanding Stripe for Non-Tech Website Builders

Understanding Stripe for Non-Tech Website Builders

Stripe is the backbone of payments on my e-commerce site, youraistudybuddy.com, where I sell *IntelliGloss: The Definitive AI Dictionary* across 20+ countries and power subscriptions with my AI Study Buddy chatbot guiding users. As a non-technical entrepreneur, I found Stripe's seamless integration and clear setup a lifesaver, though linking it to pages like my $9.95/month Starter Plan took patience due to WordPress navigation quirks. This chapter explains who Stripe is, why it's a top payment gateway, and how to use it to build a global store without coding. From setting up subscriptions to troubleshooting errors, you'll learn to make Stripe work for your brand, just as it does for mine.

Who Is Stripe and What It Does

Stripe is a leading online payment processing platform founded in 2010 by brothers Patrick and John Collison, designed to simplify transactions for businesses of all sizes. It powers payments for millions of companies, from startups to giants like Amazon, by securely handling credit card, digital wallet, and bank transactions in 135+ currencies. For youraistudybuddy.com, I use Stripe to process *IntelliGloss* subscription payments, enabling customers in 20+ countries to pay with cards,

Apple Pay, or local methods like Alipay. Its developer-friendly tools and beginner-accessible plugins (e.g., WP Simple Pay, WooCommerce) make it ideal for non-technical users building e-commerce sites on WordPress or Hostinger.

Why Stripe Is a Critical Payment Gateway

Stripe stands out as a payment gateway—a service that connects your site to banks for secure transactions—due to its reliability, global reach, and ease of use. It processes over $1 trillion annually, powering 2% of global GDP, and supports businesses in 46 countries, making it perfect for my *IntelliGloss* sales across diverse markets. Unlike traditional gateways with complex setups, Stripe integrates seamlessly with WordPress via plugins, requiring no coding to handle one-time or recurring payments, like my $19.95/month Family Basic plan. Its PCI-compliant security and fraud detection protect customers, while its dashboard provides real-time analytics, helping you track sales without the navigation headaches I faced with Hostinger.

Benefits of Using Stripe for Your E-commerce Site

Stripe's benefits make it a go-to for non-technical builders like me. It supports 135+ currencies and 100+ payment methods, enabling global sales for *IntelliGloss* without manual conversions. Its subscription tools power my Starter and Family plans, with features like free trials and automatic retries for failed payments, reducing churn. Stripe's 2.9% + $0.30 per transaction fee (no monthly costs) is transparent, and its dashboard tracks metrics like Monthly Recurring Revenue (MRR), which I monitor for youraistudybuddy.com. Plugins like WP Simple Pay simplify setup, and AI integrations (e.g., my AI Study Buddy linking to payment pages) boost conversions, making Stripe a versatile, cost-effective choice. *See Glossary: Recurring Billing, MRR.*

Setting Up Stripe for Subscriptions

Setting up Stripe for subscriptions, like my $9.95/month Starter Plan, is straightforward with WordPress. First, create a Stripe account at `dashboard.stripe.com/register`, verify your business, and connect your bank. In WordPress, install WP Simple Pay (**Plugins → Add New**), go to **WP Simple Pay → Settings → Stripe**, and enter your API keys from Stripe's **Developers → API Keys**. Create a subscription product (**Stripe → Products → Create Product**), set prices (e.g., $19.95/month for Family Basic), and enable trials (e.g., 7 days). In WP Simple Pay, add a subscription form (**Add New → Subscription Form**), link it to your Stripe product, and embed the shortcode (e.g., `[simpay id="90"]`) on a page like `youraistudybuddy.com/family-basic-payment`. Link buttons on your "Choose Your Plan" page to these URLs, as I did after fixing redirect issues. *See Glossary: Subscription Plan, Payment Form.*

Connecting Stripe to Your WordPress Pages

Properly linking Stripe payment forms to your site's pages ensures a smooth user experience, a lesson I learned when my Starter Plan button misfired. Create dedicated pages (**Pages → Add New**) for each plan, embedding shortcodes like `[simpay id="89"]` for my Starter Plan. In **Appearance → Menus**, add links to pages like `youraistudybuddy.com/starter-plan-payment` to your navigation menu. On your "Choose Your Plan" page, edit buttons (**Edit Page**) to point to the correct URLs (e.g., "Start Family Premium" to `youraistudybuddy.com/family-`

`premium-payment`). Test links in an incognito window to avoid caching errors, and clear Hostinger's cache (**hPanel → Cache → Purge All**) if redirects persist, as they did on my site. This keeps *IntelliGloss* buyers engaged without confusion.

Linking Stripe to Subscription Pages

Connecting Stripe to your pages prevents user frustration, like when my Starter Plan button led to the wrong page. Create dedicated pages (**Pages → Add New**) for each plan, adding shortcodes like `[simpay id="89"]` for my Starter Plan or `[simpay id="90"]` for Family Basic. In **Appearance → Menus**, add these pages (e.g., `youraistudybuddy.com/family-premium-payment`) to your navigation. On your "Choose Your Plan" page, edit buttons (**Edit Page**) to link to the correct URLs, ensuring "Start Family Basic" points to its payment page. Test links in an incognito window to catch errors, and clear Hostinger's cache (**hPanel → Cache → Purge All**) if redirects persist, as I did for *IntelliGloss* checkouts. *See Glossary: Redirect.*

Integrating Stripe with Your AI Chatbot

Stripe pairs powerfully with my Botpress-powered AI Study Buddy, guiding users to *IntelliGloss* subscriptions. In Botpress Studio, create flows (**Flows → New Flow**) to answer queries like "What's the Starter Plan?" with links to `youraistudybuddy.com/starter-plan-payment`. Use Stripe webhooks (**Developers → Webhooks**) to trigger post-payment actions, like sending thank-you emails. Test the flow: visit your site, ask the chatbot about plans, and confirm it directs to the correct payment page. This integration boosted my conversions by making checkouts intuitive, especially for global users. If links fail, recheck Botpress URLs and WordPress shortcodes, ensuring a seamless experience. *See Glossary: AI Agent, Webhook.*

Troubleshooting Stripe Setup Challenges

Stripe issues, like my mislinked payment forms, are common but fixable. If your form doesn't load, verify the shortcode in WordPress and check **WP Simple Pay → Settings → Stripe** for valid API keys. If payments fail, review Stripe's dashboard (**Payments → All Payments**) for errors like invalid cards. Test subscriptions in Stripe's Test Mode (card: `4242 4242 4242 4242`) to avoid real charges, as I did for my $9.95/month plan. If redirects occur, reset WordPress permalinks (**Settings → Permalinks → Save Changes**) or clear caches. Stripe's 24/7 support and WP Simple Pay's helpdesk resolved my form errors fast, keeping *IntelliGloss* sales on track.

Optimizing Stripe for Global Customers

Stripe's global features are vital for stores like youraistudybuddy.com, serving 20+ countries. Enable local payment methods (**Settings → Payment Methods**) like iDEAL for the Netherlands or UPI for India to reduce cart abandonment. Set dynamic currency conversion to show prices in local currencies, as I did for *IntelliGloss* buyers in Asia. Add tax rules (**Tax → Tax Rates**) for compliance in markets like Europe, ensuring legal sales. Use Stripe's analytics (**Reports → Overview**) to monitor Customer Lifetime Value (CLV) and optimize plans like my Family Premium tier. These tools make your store feel local to every customer, enhancing trust and sales. *See Glossary: CLV.*

Keeping Stripe Secure and Up-to-Date

Stripe's PCI-compliant security protects your customers, but maintenance is key. Update WordPress plugins like WP Simple Pay (**Dashboard → Updates**) to prevent payment glitches, as I learned when an outdated plugin crashed my subscription page. Enable Stripe Radar (**Settings → Radar**) to block fraudulent transactions, crucial for my global audience. Back up your site via Hostinger (**hPanel → Backups**) before updates to avoid data loss. Check Stripe's changelog for features like subscription discounts, keeping your *IntelliGloss* checkouts competitive. Regular maintenance ensures a secure, reliable payment system that customers trust.

Tips for Non-Technical Success with Stripe

Stripe's simplicity transformed my e-commerce journey, but its WordPress integration required patience. Start with WP Simple Pay for quick setup, test forms in incognito mode, and monitor sales in Stripe's dashboard. Use free resources like Stripe's guides or WPBeginner to troubleshoot. For youraistudybuddy.com, I focused on one task—linking my Starter Plan—before scaling to Family plans. Test every change as a customer to catch errors, like my redirect woes. Your payment system doesn't need to be flawless at first; prioritize secure transactions, then refine to align with your brand, whether selling books like *IntelliGloss* or courses globally.

Why Stripe Is a Must-Have for E-commerce Success

Stripe's reliability, global reach, and no-code integrations make it indispensable for non-technical builders like me, who built a thriving payment system for youraistudybuddy.com. Its ability to process subscriptions, integrate with AI chatbots, and support 135+ currencies ensures your store scales effortlessly. My journey—from AT&T and flower shops to an AI education leader—shows that Stripe's tools, despite setup challenges, empower anyone to create a professional checkout experience. Take it one step at a time, test diligently, and you'll build a payment system that drives sales worldwide, just like *IntelliGloss* does for me.

Areas to Strengthen or Add

1. Pricing Transparency

1. No visible pricing details or package breakdown. Including clear price tiers (e.g., hourly, program lengths) builds trust and avoids friction at signup.

2. Detailed Tutor Profiles

1. The "Team" section likely introduces tutors, but lacks detail on credentials, areas of expertise, teaching style, or sample tutor bios. Potential clients want to see who they're working with.

3. Success Metrics & Case Studies

1. Testimonials are generic; consider adding:
    1. Data examples ("Jane improved from 70% to 90% math scores").

    2. In-depth case studies or parent/student interviews.

4. Updated Content in Blog/FAQs

1. The blog appears present, but check for freshness and relevance (last posts etc.). Regular, valuable posts can help SEO and show your knowledge edge.
2. FAQs might benefit from expansion: session structure, cancellations, progress tracking, tech requirements.

5. Technical SEO & Usability

1. Ensure site is mobile-responsive, and loads quickly (especially for parents browsing on phones).
2. Technical aspects like meta tags, structured data for local business, schema for tutors, and SSL certificate are critical.

6. Analytics & Tracking

1. No visible mention of analytics (e.g., Google Analytics, Tag Manager), conversion tracking, or heatmaps. These help optimize CTAs, ad campaigns, and content strategy.

7. Trust & Social Proof

1. Consider more:
    1. Third-party verification (Google/Trustpilot reviews).
    2. Press mentions.
    3. Credentials such as tutor qualifications, background checks, or partnerships with schools.

8. Interactive Features

1. Features that can stand out include:
    1. Live chat or automated scheduling tools.
    2. Free assessment quizzes or trial session booking.
    3. Parent portal demo to track student progress.

9. Privacy Policy & Terms

1. While there's a footer link, make sure privacy policy, especially concerning underage user data, is robust and compliant.

10. Local SEO & Reach

1. For in-person tutoring, add:
    1. Maps showing service areas or tutor locations in Los Angeles (if applicable).
    2. Local landing pages for specific neighborhoods or subjects.

✅ Quick Action Suggestions

Area	Action
Pricing	Add tiered plans, session bundles, trial pricing
Tutors	Create profiler pages with credentials, specialties
Stories	Publish case studies and quantifiable success stories
Content	Maintain active blog & expanded FAQs
Tech	Confirm mobile-responsiveness & install analytics
Trust	Integrate third-party review platforms, badges
Tools	Introduce live chat, online booking, free assessments
Locality	Expand local SEO, maps, and localized content

☑ To Do Next

1. **Perform a content audit**: Check if blog/FAQs are current.
2. **Technical test**: Run Google PageSpeed and mobile usability audit.
3. **Survey your users**: Ask for quantifiable outcomes, get reviewers.
4. **Enhance SEO**: Optimize for tutors and tutoring-related queries in LA.

By addressing these aspects, **Your AI Study Buddy** can present a much stronger brand—more transparent, trustworthy, and optimized for conversions.

Enhancing Engagement with AI and Stripe

Stripe pairs beautifully with AI tools like my Botpress-powered AI Study Buddy, which guides users to *IntelliGloss* subscription pages. In Botpress, create flows for queries like "What's the Family Basic plan?" with responses linking to `youraistudybuddy.com/family-basic-payment`. In Stripe, use webhooks (**Developers → Webhooks**) to trigger actions (e.g., email confirmations after payments), enhancing automation. For example, my chatbot promotes my YouTube channel and subscription plans, boosting conversions. Test these integrations to ensure they align with your brand's tone, making checkouts as seamless as my global *IntelliGloss* sales. *See Glossary: AI Agent.*

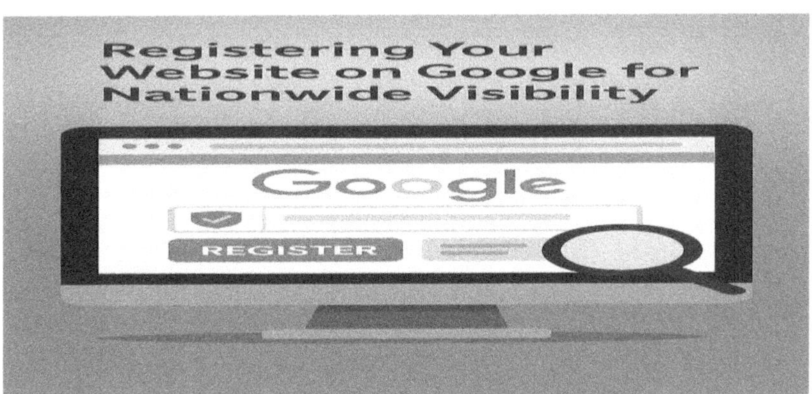

Chapter 60: Registering Your Website on Google for Nationwide Visibility

Making **youraistudybuddy.com** visible on Google changed how I could reach teachers, students, parents, and everyday people searching for guides on building websites, e-commerce sites, chatbots, and educational products like *IntelliGloss: The Definitive AI Dictionary*. As a non-technical entrepreneur, I had navigation challenges with tools like WordPress and Hostinger, but registering my site on Google was straightforward and helped people in 20+ countries find it. This chapter gives the step-by-step process I used to register a site on Google, optimize for searches like "how to build an e-commerce site" or "educational products," and improve nationwide visibility. With tools like **Google Search Console** and **Google Business Profile**, you can make your site discoverable, drive qualified traffic, and grow your brand — even without coding skills.

Why Registering on Google Is Crucial for Your E-commerce Site

Registering your website on Google helps ensure it appears in search results for terms like "how to build a website" or "educational products." For **youraistudybuddy.com**, this meant teachers and students could find my *IntelliGloss* pages and the AI Study Buddy chatbot. Google crawls billions of pages; new sites or sites with few backlinks can be missed. **Google Search Console (GSC)** helps Google index your pages, while **Google Business Profile (GBP)** improves local visibility for schools and parents. Together, these tools can increase relevant traffic, improve conversions (for example, to a starter subscription plan), and build trust. *See Glossary: SEO.*

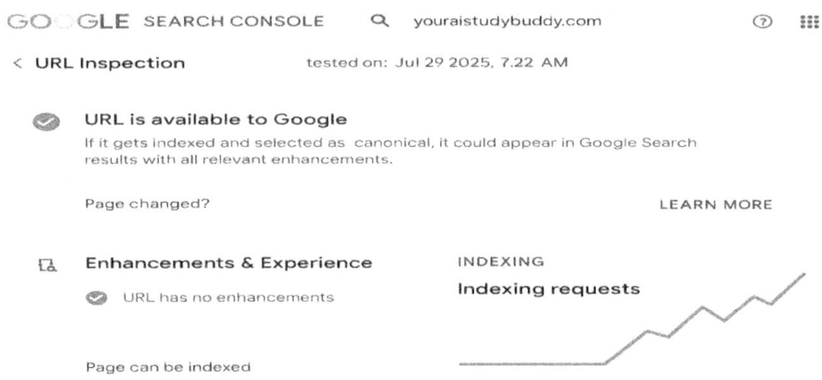

Step 1: Set Up Google Search Console for Indexing

Google Search Console (GSC) is a free tool that helps Google find and index your site so it can appear in search results.

1. **Create a Google Account:** Go to `accounts.google.com`, sign up or log in, ideally with an address like `info@youraistudybuddy.com`.

2. **Access GSC:** Visit `search.google.com/search-console`, click **Start Now**, and choose **Add Property**.
3. **Add Your Website:** Choose **URL Prefix**, enter `https://youraistudybuddy.com` (make sure it matches your SSL setup in Hostinger), and click **Continue**.
4. **Verify Ownership:**
   - **HTML file:** Download the verification file from GSC, upload it to **hPanel → File Manager → public_html**, then click **Verify**.
   - **OR HTML tag:** Copy the meta tag and paste it into your WordPress header (**Appearance → Theme File Editor → header.php**).
5. **Check Indexing:** In GSC, open **Pages → Why pages aren't indexed** and confirm key URLs (e.g., `youraistudybuddy.com/starter-plan-payment`) are indexed. If not, continue to Step 3.

This process took me about 10 minutes. *See Glossary: Google Search Console.*

Step 2: Submit a Sitemap to Google

A **sitemap** lists your pages (for example, "Choose Your Plan" or chatbot pages) so Google can find them efficiently.

1. **Install Yoast SEO:** In WordPress, **Plugins → Add New → Yoast SEO → Activate**.
2. **Generate Sitemap:** Yoast creates one at `youraistudybuddy.com/sitemap_index.xml`. Visit that URL to confirm it loads.
3. **Submit to GSC:** In GSC, **Sitemaps → Add a new sitemap**, enter `sitemap_index.xml`, and click **Submit**.
4. **Request Indexing for New Pages:** In **URL Inspection**, enter a new page (like a new IntelliGloss product URL) and click **Request Indexing**.

If you're using Hostinger Website Builder, create a sitemap in **Website Builder → Settings → SEO** and submit it the same way. *See Glossary: Sitemap.*

Step 3: Optimize Your Site for Relevant Searches

Basic SEO helps you appear for terms like "how to build an e-commerce site" or "educational products."

1. **Research Keywords:** Use **Google Keyword Planner** to find terms like "build a website for beginners."
2. **Add Keywords to Content:** Update page titles, headings, and body copy (e.g., "Build Your E-commerce Site with Your AI Study Buddy").
3. **Optimize Meta Tags:** In **Yoast → Search Appearance**, set concise titles (≈60 chars) and meta descriptions (≈160 chars).
4. **Use Alt Text:** Add descriptive alt text to images (e.g., "IntelliGloss AI Dictionary book cover").
5. **Improve Site Speed:** In Hostinger, enable **LiteSpeed Cache** (**hPanel → Websites → Manage → Cache**).

These steps helped **youraistudybuddy.com** show for "how to build a chatbot"-type searches. *See Glossary: Meta Tag.*

Step 4: Set Up a Google Business Profile for Local Visibility

A **Google Business Profile** (GBP) can increase visibility in local and map results.

1. **Create a Profile:** Go to `business.google.com`, sign in, and click **Create Profile**. Enter your business name (e.g., "Your AI Study Buddy").
2. **Add Details:** Include your website, phone, service area (e.g., United States), and categories like **Educational Consultant** or **E-commerce Service**.
3. **Verify:** Complete postcard or phone verification.
4. **Optimize:** Add a clear description, photos, and links to important pages (e.g., `youraistudybuddy.com/starter-plan-payment`). Post updates about new resources or tutorials.
5. **Encourage Reviews:** Ask customers to leave reviews to build credibility.

This helped my site show in Google Maps and local searches for schools and parents. *See Glossary: Google Business Profile.*

Step 5: Integrate with Educational Platforms

To connect with classrooms and educators:

1. **Explore Google for Education:** See `edu.google.com` for tools like **Google Classroom**. Offer a free PDF resource on your blog to attract teachers.
2. **Create Education-Focused Pages:** For example, "Educational Products for Teachers," with keywords like "classroom resources."
3. **Earn Backlinks:** Share useful guides on legitimate education forums or communities and link back to your site.

These steps increased teacher interest and relevant traffic. *See Glossary: Backlink.*

Step 6: Enhance Chatbot and E-commerce Visibility

Make sure your chatbot and product pages can be discovered.

1. **Chatbot Page SEO:** Create a page like **How to Build a Chatbot**, embed your Botpress widget, and include beginner-friendly keywords. Add a call-to-action to your subscription form (e.g., `[simpay id="89"]`).
2. **E-commerce SEO:** Optimize product pages (e.g., your shop) with clear titles, descriptions, and internal links.
3. **Connect Botpress to Payments:** In Botpress flows, direct users to relevant payment pages (e.g., `youraistudybuddy.com/family-basic-payment`) so the path is clear.

I set up these flows to make checkouts straightforward. *See Glossary: AI Chatbot.*

Step 7: Monitor and Maintain Visibility

Keep an eye on performance and make small, regular improvements.

1. **Track Queries:** In GSC **Performance → Search results**, see which terms drive clicks and refine pages accordingly.
2. **Fix Indexing Issues:** Use **URL Inspection** to identify unindexed pages and request indexing.
3. **Update Content:** Publish periodic posts (e.g., "Top 5 Tips for Non-Tech Website Builders").
4. **Maintain GBP:** Post updates occasionally and keep details accurate.

This ongoing process helped me see where to focus next. *See Glossary: Indexing.*

Troubleshooting Common Google Registration Issues

If your site isn't appearing as expected:

1. **Not Indexed:** In GSC **Pages**, look for "Blocked by robots.txt." In **hPanel → File Manager**, adjust `robots.txt` to allow crawling:

   ```
   makefile
   CopyEdit
   User-agent: *
   Allow: /
   ```

2. **Redirect Errors:** Clear **Hostinger cache** (**hPanel → Cache → Purge All**) and reset WordPress permalinks (**Settings → Permalinks → Save Changes**).
3. **Low Rankings:** Strengthen content and earn a few quality backlinks.
4. **GBP Delays:** If a postcard doesn't arrive, contact **support.google.com/business**.

Always test pages in an incognito window to see what visitors see. *See Glossary: Redirect.*

Tips for Non-Technical Success with Google

Start with GSC and GBP for quick wins, test like a visitor, and use free references such as Google's **SEO Starter Guide** or **WPBeginner**. I limited plugins to avoid slowdowns, and I focused on one improvement at a time (for example, submitting the sitemap first). Rankings build over time; prioritize indexing and clear, helpful content.

Why Google Registration Supports E-commerce Growth

Registering your site with Google and applying a few basics in SEO and GBP helps the right people discover your content, courses, or products. In my experience, working step by step — from verification and sitemaps to small SEO updates — made a noticeable difference in visibility. Take it steadily, keep testing, and let the improvements compound.

Chapter 61: Promoting Your E-commerce Site on Google with Ads and SEO

Promoting youraistudybuddy.com on Google allowed me to connect with teachers, students, parents, and schools who were looking for website-building guides, chatbot tutorials, and educational resources. As someone who had previously worked with Google Ads (back when it was called Google AdWords) at AT&T, I understood the potential of these tools — but after years away, I found myself having to relearn everything while also navigating the quirks of WordPress.

This chapter walks you through the practical steps I took to set up and run Google Ads campaigns and apply SEO strategies to increase visibility. My goal is to show how someone without advanced coding or marketing skills can still use these tools to reach a wider audience.

Why Google Ads and SEO Are Important for E-commerce

Google Ads and SEO form a powerful combination for online visibility. Ads allow you to appear immediately in search results for targeted keywords, while SEO builds your presence in organic rankings over time. Together, they help your website be found by people actively searching for the type of content, products, or services you offer.

For me, using both meant my site was more discoverable for search terms like "how to build an e-commerce site," "educational products," or "build a chatbot." PPC (pay-per-click) ads provided targeted exposure, while SEO ensured that the same topics could also bring in visitors naturally.

Step 1: Set Up a Google Ads Account

1. Go to ads.google.com and sign in with your Google account (use a professional email like info@youraistudybuddy.com).
2. Enter your website details, set your location (for example, United States), and add billing information.
3. Choose your main objective (for me, it was "Sales" to encourage purchases and subscriptions).
4. Familiarize yourself with the campaign types: Search (text ads), Shopping (product ads), Display (visual banners), and Video (YouTube).

Step 2: Launch PPC Search Ads for Targeted Traffic

Search Ads are triggered when someone searches for keywords you choose.

1. Create a campaign under **Sales → Search Campaign** and select "Website visits."
2. Set a starting budget (I began with $10–$20/day).
3. Use Keyword Planner to find phrases relevant to your audience.

4. Write clear ad headlines and descriptions with a call-to-action (e.g., "Learn to Build Your Website").
5. Include sitelinks to key pages like your "Plans" or "Shop" section.

Step 3: Set Up Shopping Ads for Products

If you sell physical or digital products, Shopping Ads can display them directly in Google's shopping results.

1. Create a Merchant Center account and connect it to your site.
2. Upload your product details (title, description, price, and images).
3. Launch a Shopping Campaign and set a daily budget.
4. Keep product titles clear and keyword-focused.

Step 4: Create Display Ads for Awareness

Display Ads are banners that appear across websites in Google's network.

1. Choose your audience based on interests, topics, or demographics.
2. Design simple, eye-catching banners using a tool like Canva.
3. Link the ads to relevant pages on your site.

Step 5: Use Video Ads for Engagement

Short videos can quickly communicate your message.

1. Record a short clip (15–30 seconds) introducing your service or tutorial.
2. Upload it to YouTube and run it as an ad targeting your chosen audience.

Step 6: Apply SEO for Long-Term Results

1. Research keywords with moderate competition and good search volume.
2. Add these keywords naturally to your site's pages, blog posts, and product descriptions.
3. Use a plugin like Yoast SEO to manage titles, descriptions, and readability.
4. Build backlinks by contributing to relevant blogs or forums.
5. Keep your site fast and mobile-friendly.

Step 7: Integrate Ads with Your Chatbot

If you have a chatbot, you can send ad traffic directly to it.

1. Create chatbot flows that guide visitors to useful information or purchasing options.
2. Link your ads to the chatbot's landing page.
3. Monitor engagement and refine the chatbot's responses over time.

Step 8: Monitor and Optimize

1. Check your ad performance regularly in Google Ads.
2. Use Google Search Console to track how your site is performing organically.

3. Test different ad headlines, keywords, and landing pages to improve results.

Troubleshooting Common Issues

- If your ads are disapproved, review Google's ad policies.
- If rankings are low, improve your site's content quality and backlink profile.
- If your site is slow, optimize images and enable caching.

Final Thoughts
Using Google Ads alongside SEO is not about overnight results — it's about building consistent visibility and attracting the right visitors to your site. Whether you sell educational tools, offer online courses, or run a niche service, these steps can help you create a balanced strategy that works over time.

Epilogue: Your WordPress Journey Starts Here

When I first opened my laptop to start building a website, I didn't have a clue about WordPress menus, hosting plans, or the difference between a plugin and a widget. I wasn't a web developer. I wasn't a coder. I was simply a business owner with a dream — and the determination to make it happen.

If you've made it to this point in the book, you now have something even more powerful than technical skills: **a roadmap**. You've learned how to choose your hosting, install WordPress, create pages that tell your story, and connect payment gateways that can serve customers worldwide. You've discovered that you can integrate tools like chatbots, SEO, and security features without needing a computer science degree.

The truth is, building a website is less about being "tech-savvy" and more about being **resourceful, consistent, and willing to learn**. Every button you click, every error you troubleshoot, every plugin you try — it's all part of your growth.

Remember, websites are not "set it and forget it" projects. They are living, evolving spaces. Keep your content fresh. Update your plugins. Listen to your visitors. Make small changes, test them, and improve as you go. The beauty of WordPress is that it grows with you — from your first page to your first 1,000 customers.

Most importantly, don't let the fear of technology hold you back. I've proven that even without a tech background, you can create a website that works, looks professional, and earns income. If I can do it, so can you.

This is just the beginning. Your WordPress journey doesn't end here — it starts here. So take what you've learned, launch your site, and let the world see what you have to offer.

Glossary of E-commerce Terms

This glossary explains key terms you'll encounter while building an e-commerce website like **youraistudybuddy.com**, which I created to sell *IntelliGloss: The Definitive AI Dictionary* and share my YouTube videos. As a non-technical entrepreneur, I know how confusing jargon can be—terms like "DNS" or "API" frustrated me as much as WordPress's hidden menus.

I designed this glossary to serve as a **quick reference guide**, helping you master the tools and concepts that power your online business—from setting up WordPress to connecting Stripe. Along the way, you'll also see how **multi-agents**—specialized AI assistants that can work together—are transforming e-commerce by handling different tasks like customer support, product recommendations, and order management simultaneously.

Use this resource as a bridge between technical language and real-world application, so you can focus less on decoding jargon and more on building a store that works for you.

GLOSSARY:

LETTER A

Access Control
A security process that determines who can view or use resources on a website. In WordPress, it involves setting permissions for different user roles such as Administrator, Editor, Author, and Subscriber.

Accessibility
The practice of designing and developing websites so that people with disabilities can use them. This includes adding features like screen-reader support, alt text for images, and high-contrast colors.

Account Settings
A section in WordPress (often in the user profile area) where you can manage your personal information, password, display name, and preferences.

Action Hook
A WordPress programming feature that lets developers add custom functions at specific points during WordPress's execution. Action hooks make it possible to modify or extend site functionality without editing core files.

Active Theme
The theme currently applied to your WordPress site that controls the layout, design, and appearance.

Admin Area
The back-end dashboard of WordPress where site owners and editors manage content, settings, plugins, and themes.

Admin Bar
The horizontal toolbar that appears at the top of the WordPress interface (both in the dashboard and on the front-end for logged-in users) providing quick access to admin functions.

Admin Dashboard

Definition
The Admin Dashboard is the central control panel in WordPress where users manage content, settings, themes, plugins, and other features of the website.

Key Components

1. Left-hand navigation menu
2. Quick access widgets (e.g., At a Glance, Activity)
3. Toolbar for updates, users, and notifications

Benefits

1. Streamlined control over website settings
2. Easy access to editing tools and configurations
3. Helps manage users, updates, and performance

Applications
Ideal for bloggers, developers, and business owners to monitor and manage their WordPress sites.

Industry
Web development, digital content, small businesses

Challenges

1. Can be overwhelming for beginners
2. Poor layout from plugin overload may cause confusion

Summary
The Admin Dashboard serves as the website's control tower, helping users manage and operate their WordPress site from one convenient interface.

Admin Panel
Another term for the WordPress admin area where all management and configuration tasks are performed.

Affiliate Link
A special URL containing a tracking code that earns a commission when visitors click it and make a purchase. Often used in affiliate marketing programs.

AGI (Artificial General Intelligence):
A type of artificial intelligence designed to perform any intellectual task that a human can do. Unlike narrow AI, which is built for specific jobs (like translation, chess, or image recognition), AGI would have flexible, all-around intelligence — able to learn new skills, adapt to unfamiliar problems, and apply knowledge across different fields. It represents the idea of machines reaching human-like adaptability and problem-solving ability.

AI (Artificial Intelligence):
A branch of computer science that creates systems able to perform tasks that usually require human intelligence, such as learning, problem-solving, recognizing patterns, and understanding language. AI does not "think" like a person, but it uses algorithms and data to simulate aspects of human decision-making. Examples include voice assistants, image recognition, chatbots, and recommendation systems.

AI Agents

Definition

An AI agent is a software program or system that can perceive its environment, process information, and take actions to achieve specific goals. It uses artificial intelligence techniques—such as natural language processing, decision-making, or machine learning—to interact with users, other agents, or digital platforms.

Key Components

- **Perception**: The ability to gather input (e.g., text, speech, or data).
- **Reasoning/Decision-Making**: Processing input and selecting the best response or action.
- **Action**: Carrying out tasks, like answering questions, sending emails, or triggering workflows.
- **Learning**: Improving performance over time by analyzing past interactions.

Examples

- A chatbot on an e-commerce site that answers customer questions.
- A virtual assistant that schedules meetings or manages emails.
- A recommendation agent that suggests products based on browsing history.

Benefits

- Automates repetitive tasks.
- Provides 24/7 support.
- Improves personalization for users.
- Can work alongside human staff to increase efficiency.

Applications

- E-commerce customer service.
- Healthcare (appointment reminders, symptom checkers).
- Finance (fraud detection, budgeting tools).
- Education (tutoring, study aids).

Industries

- Retail
- Banking and FinTech
- Healthcare
- Education and Training

Challenges

- May produce errors if not trained properly.
- Requires safeguards to handle sensitive data securely.
- Needs continuous updates to remain accurate and effective.

Summary
AI agents act as intelligent digital assistants that perceive, reason, and act to achieve goals. They are central to modern applications like chatbots, recommendation systems, and virtual assistants, making technology more responsive and user-friendly.

AI Chatbot
An artificial intelligence program integrated into a website that can simulate conversation, answer questions, and help visitors navigate the site.

AI Multi-Agents

Definition
AI multi-agents (or multi-agent systems) are groups of intelligent agents that work together, communicate, and coordinate to achieve goals that are too complex for a single agent to handle. Each agent operates independently but contributes to the overall system, making the collective more powerful and adaptive.

Key Components

- **Multiple Agents**: Each agent has its own role, such as data analysis, customer support, or recommendation.
- **Communication**: Agents share information to stay updated and avoid duplication of tasks.
- **Coordination**: Agents divide responsibilities and adjust their actions to work efficiently together.
- **Autonomy**: Each agent can make decisions without waiting for direct control from a human.

Examples

- An e-commerce platform where one agent handles product searches, another manages payments, and a third provides customer support—all working in sync.
- Smart city systems where traffic agents, energy agents, and security agents collaborate to manage urban resources.

Benefits

- Handles complex tasks by breaking them into smaller, manageable parts.
- Increases efficiency through parallel work.
- More resilient—if one agent fails, others can continue operating.
- Supports scalability as new agents can be added for new tasks.

Applications

- **E-commerce**: Order processing, fraud detection, and chat support managed by different agents.
- **Robotics**: Swarms of robots collaborating on construction or search-and-rescue missions.
- **Finance**: Multiple agents monitoring transactions, compliance, and customer inquiries simultaneously.
- **Healthcare**: Agents coordinating patient data, scheduling, and diagnostic support.

Industries

- Retail and online platforms
- Transportation and logistics
- Banking and FinTech
- Healthcare and smart hospitals
- Education (multi-bot tutoring systems)

Challenges

- Ensuring smooth communication among agents.
- Avoiding conflicts when agents make competing decisions.
- Managing system complexity as more agents are added.
- Security risks if malicious agents enter the network.

Summary
AI multi-agents represent the next step beyond single AI assistants. By working together as a team, these systems can solve complex, large-scale problems more effectively. They are becoming increasingly important in e-commerce, finance, robotics, and smart infrastructure.

AJAX
(Asynchronous JavaScript and XML) – A technology that allows web pages to send and receive data from a server without reloading the page. Used in WordPress for features like live search or instant updates.

Alt Text
Short for "alternative text." A description of an image that is displayed if the image can't load, and read aloud by screen readers for accessibility.

Anchor Text
The clickable text in a hyperlink. For example, in <a href="https://example.com">Learn More</a>, the words "Learn More" are the anchor text.

API
(Application Programming Interface) – A set of rules and tools that allows different software applications to communicate with each other. WordPress has APIs for plugins, themes, and external services.

Appearance Menu
A section in the WordPress admin area where you can customize your site's look by managing themes, widgets, menus, and background settings.

Archive Page
A WordPress page that lists multiple posts grouped by category, tag, date, or author.

Attachment Page
A single page generated by WordPress for each uploaded media file, displaying that file along with optional metadata.

Audio File
A sound file (such as MP3, WAV, or OGG) that can be uploaded and played on a WordPress site.

Author Box
A small section, usually at the end of a post, that shows information about the post's author such as their name, bio, and profile picture.

Author Profile
A page or section displaying an author's information, often including their bio, links to their website or social media, and a list of their posts.

Author URL
The web address associated with an author's profile, which may lead to their personal website or profile page.

Auto-Update
A WordPress feature that automatically installs updates for themes, plugins, or the WordPress core without manual intervention.

Ad Group: A set of ads in Google Ads targeting specific keywords. My *IntelliGloss* ad group targets "build a website," driving focused traffic. *See Google Ads.*

AI Agent
Definition: An AI-powered software that performs tasks or interacts with users autonomously, like a chatbot.

Explanation: On youraistudybuddy.com, my AI Study Buddy chatbot uses Botpress to answer questions about *IntelliGloss* and guide users to subscription pages like `[simpay id="89"]`. AI agents work 24/7, enhancing customer engagement without human intervention, crucial for scaling e-commerce globally. They learn from user interactions to improve responses.
Cross-reference: AI Chatbot, Botpress, NLP. See "Understanding Botpress."

Key Components

- **Perception:** The ability to collect and process data from the environment (e.g., sensors, text, voice, or online data).
- **Reasoning/Decision-Making:** Using algorithms or models to analyze data, predict outcomes, and decide on actions.
- **Action:** Carrying out tasks such as responding to users, executing commands, or controlling devices.
- **Learning:** Improving performance over time through feedback and new data.

Benefits

- Automates repetitive or complex tasks.
- Enhances efficiency and productivity.
- Provides personalized recommendations and support.
- Can operate continuously without fatigue.

Applications

- **Virtual Assistants:** Siri, Alexa, or ChatGPT-like bots.
- **E-commerce:** Recommending products or managing customer service.
- **Robotics:** Controlling autonomous vehicles, drones, or household robots.
- **Healthcare:** Monitoring patients or assisting in diagnosis.

Summary
An AI Agent is essentially an intelligent "helper" that senses, thinks, and acts to complete tasks or solve problems. It can range from simple chatbots to advanced autonomous robots, making it a key part of modern AI systems.

AI Chatbot
Definition: A conversational program using artificial intelligence to interact with users via text or voice.
Explanation: My AI Study Buddy on youraistudybuddy.com answers queries like "What's the Starter Plan?" and links to payment pages, boosting *IntelliGloss* sales. Built with Botpress, it handles multiple languages for 20+ countries, saving time and reducing support costs. Non-technical users can create chatbots without coding.
Cross-reference: AI Agent, Botpress, Webhook. See "Understanding Botpress."

Analytics
Definition: Tools to track website performance and user behavior.
Explanation: Google Analytics tracks youraistudybuddy.com's traffic, showing how many visit *IntelliGloss* pages from "build a chatbot" searches. Insights help optimize ad campaigns for

nationwide reach.
Cross-reference: Google Ads, Conversion Rate. See "Promoting Your E-commerce Site."

API Key
Definition: A unique code that allows two software systems to communicate securely.
Explanation: In Stripe, I used API keys to connect WP Simple Pay to my youraistudybuddy.com payment forms, enabling secure transactions for *IntelliGloss* subscriptions. Found in Stripe's dashboard, API keys ensure data privacy and must be kept confidential to prevent unauthorized access.
Cross-reference: Payment Gateway, Stripe. See "Understanding Stripe."

ASCII (American Standard Code for Information Interchange)

Definition
ASCII is a character encoding standard that assigns numeric values to letters, digits, punctuation marks, and control characters. It allows computers and electronic devices to represent and exchange text in a universal format.

Key Components

- **Character Set:** Consists of 128 characters, including uppercase and lowercase English letters (A–Z, a–z), digits (0–9), punctuation marks, and control codes (like newline or tab).
- **Encoding Scheme:** Each character is represented by a 7-bit binary number (values 0–127). For example, the capital letter **A** is represented by the number **65** in ASCII.
- **Control Characters:** Special codes that manage text formatting and device functions, such as carriage return or end of transmission.

Benefits

- Provides a standard way for computers to communicate text data.
- Simple and widely supported across hardware and software.
- Forms the foundation for many modern encoding systems like UTF-8.

Applications

- Data exchange between computers and devices.
- Text files, programming, and internet protocols.
- Early email and network communications.

Summary
ASCII is one of the earliest and most influential text encoding standards. Though limited to 128 characters, it laid the groundwork for modern encodings and remains a fundamental concept in computing today.

LETTER B

Backlink
Definition: A link from another website pointing to your site, boosting SEO.
Explanation: I shared *IntelliGloss* guides on educational blogs to earn backlinks, improving youraistudybuddy.com's Google ranking for "build a website." Backlinks signal trust to search engines, driving organic traffic from teachers and students. Quality matters more than quantity.
Cross-reference: SEO, Google Search Console. See "Registering Your Website on Google."

Back End
The behind-the-scenes part of a website that manages data, servers, and functionality, like processing orders or storing customer info.
Example: On youraistudybuddy.com, the back end handles *IntelliGloss* order data in WooCommerce and stores chatbot responses in Botpress, ensuring smooth transactions. *See WordPress: Plugins, Botpress: Knowledge Base.*

Backup Plugin

Definition
A Backup Plugin is a tool that automates the process of saving copies of your WordPress site's files and database.

Key Components

1. Scheduled backups
2. Cloud storage integration
3. Manual restore options

Benefits

1. Protects against data loss
2. Quick site recovery
3. Peace of mind for site owners

Examples
Plugins like UpdraftPlus, BackupBuddy, and Jetpack Backup offer automated daily backups.

Applications
Used on all types of sites to ensure business continuity.

Challenges

1. Large backups may use server space
2. Improper settings can lead to incomplete backups

Summary
A backup plugin is your digital insurance policy, securing your site's content and settings against accidental or malicious loss.

Bandwidth

The amount of data transferred between your website and visitors within a specific time period. Higher bandwidth means your site can handle more traffic and large files without slowing down.

Block

A content element in WordPress's **Block Editor (Gutenberg)**. Blocks can be paragraphs, images, buttons, videos, or other design elements you place on your page.

Block Editor

Also called **Gutenberg**, this is WordPress's modern content editor that uses blocks to create and design pages or posts without needing to code.

Blog

A regularly updated section of a website where articles, stories, news, or tips are posted in chronological order. WordPress was originally built as a blogging platform.

Blogroll

A list of recommended links to other blogs or websites, usually found in a sidebar. This feature is mostly outdated but still exists in some WordPress setups.

Bookmark

A saved shortcut in your web browser to quickly revisit a specific webpage.

Bot

An automated program that performs tasks online without human intervention. Bots can be good (like search engine crawlers) or bad (like spam bots).

Botpress

Definition: An open-source platform for building no-code AI chatbots.
Explanation: Botpress powers my AI Study Buddy, guiding users to *IntelliGloss* payment pages like `youraistudybuddy.com/starter-plan-payment`. Its drag-and-drop studio lets non-technical users create flows for customer support, saving costs and enabling 24/7 engagement across 20+ countries.
Cross-reference: AI Chatbot, NLP, Webhook. See "Understanding Botpress."

Bounce Rate: The percentage of visitors leaving after viewing one page. I reduced bounce rate on youraistudybuddy.com with engaging chatbot flows. *See Analytics.*

Breadcrumb

A navigation aid that shows visitors the path they took to reach a specific page. For example: **Home > Blog > WordPress Tips**.

Broken Link

A hyperlink that no longer works, often leading to a "404 Page Not Found" error. Broken links can hurt SEO and user experience.

Browser Cache

Temporary storage in a web browser that saves files from visited websites to load them faster on future visits.

Browser Compatibility

How well a website works across different browsers (like Chrome, Firefox, Safari, and Edge). A compatible site displays correctly on all major browsers.

Browser Extension

A small software add-on for a web browser that adds extra features, such as password managers, ad blockers, or grammar checkers.

LETTER C

Cache

Definition: Stored data to speed up website loading.
Explanation: Hostinger's LiteSpeed Cache on youraistudybuddy.com reduces load times for *IntelliGloss* checkouts, improving SEO and user experience. Clear caches to fix redirect issues.
Cross-reference: Hosting, Redirect. See "Understanding Hostinger."

Categories and Tags

Definition
Categories and tags are WordPress taxonomies used to organize and classify content for easier navigation and indexing.

Key Components

Category hierarchy (parent/child)

Tag clouds

Archive pages for both

Benefits

Helps users find related content

Boosts SEO through structured data

Enables content filtering

Examples
A blog may have categories like "AI Basics," "WordPress Tips," and tags like "Beginner," "Plugins," "Security."

Applications
Used in blogs, news websites, and resource libraries.

Challenges

Over-tagging or unclear labels can confuse users

Poor taxonomy structure can hinder SEO

Summary
Categories and tags help organize content and improve discoverability, making it easier for both users and search engines to explore your site.

Chatbot
A software program that automates conversations with users, often using AI to answer questions. *Example*: My AI Study Buddy chatbot on youraistudybuddy.com answers queries like "What is a GPU?" and links to *IntelliGloss*. *See Botpress: Flows.*

Checkout: The process of completing an online purchase. Stripe's checkout on youraistudybuddy.com ensures smooth *IntelliGloss* payments. *See Payment Form.*

Child Theme

Definition
A child theme is a theme that inherits the functionality and styling of another theme, called the parent theme, and allows for safe customization.

Key Components

1. `style.css` and `functions.php` files
2. Parent theme dependency
3. Override template files

Benefits

1. Allows custom changes without affecting updates
2. Safe experimentation zone for developers
3. Preserves original theme code

Applications
Used when customizing themes for clients or unique branding needs.

Industry
Web development, digital design, freelance services

Challenges

1. Requires basic understanding of CSS and PHP
2. If parent theme is poorly coded, child theme inherits problems

Summary
Child themes offer a secure and flexible way to customize WordPress sites while maintaining update compatibility with the parent theme.

Churn Rate

The percentage of subscribers who cancel or stop using your service over a specific period, indicating customer retention success.

Example: On youraistudybuddy.com, I monitor churn rate to see how many IntelliGloss subscribers cancel monthly, using my AI Study Buddy chatbot to engage users and reduce cancellations. See Stripe: Analytics, Botpress: Webchat Integration.

Custom Post Type

Definition
A Custom Post Type (CPT) is a content type beyond the default WordPress options (post, page) that allows tailored content structure.

Key Components

1. CPT registration (via code or plugin)
2. Custom fields and taxonomies
3. Admin menu integration

Benefits

1. Creates distinct content types (e.g., portfolios, testimonials)
2. Organizes data for specific use cases
3. Enhances flexibility and content modeling

Applications
Common in real estate listings, event calendars, directories, and product catalogs.

Industry
Ecommerce, real estate, media, education

Challenges

1. Requires technical setup or plugin use
2. May break during theme or plugin conflicts

Summary
Custom Post Types extend WordPress beyond blogs and pages, allowing users to organize and display content tailored to their needs.

Customer Lifetime Value (CLV)
The total revenue expected from a customer over their entire relationship with your business, based on their subscription payments.
Example: I calculate CLV for IntelliGloss subscribers on youraistudybuddy.com to estimate long-term profits, helping me plan YouTube content to keep users subscribed. See Stripe: Analytics.

CLV (Customer Lifetime Value)
Definition: The total revenue expected from a customer over their relationship with your business.
Explanation: For *IntelliGloss* subscribers, I calculate CLV in Stripe to estimate earnings from my Family Premium plan ($29.95/month). High CLV justifies ad spend on Google Ads, targeting teachers and parents nationwide to maximize long-term sales.
Cross-reference: Churn Rate, Google Ads. See "Understanding Stripe."

Content Management System (CMS)
Definition: Software for creating and managing website content without coding.
Explanation: WordPress, my CMS for youraistudybuddy.com, lets me build pages like "Shop IntelliGloss" and embed Stripe forms without technical skills. Its dashboard simplifies content updates, ideal for non-technical users creating e-commerce sites.
Cross-reference: WordPress, Dashboard. See "Understanding WordPress."

Conversion Rate
Definition: The percentage of visitors who complete a desired action, like purchasing.
Explanation: My AI Study Buddy boosts youraistudybuddy.com's conversion rate by guiding users to `[simpay id="89"]` for *IntelliGloss* purchases. A 3% conversion rate (3 out of 100 visitors buying) is strong; I improved mine with targeted Google Ads.
Cross-reference: Google Ads, AI Chatbot. See "Promoting Your E-commerce Site."

CPU (Central Processing Unit)

Definition
The CPU, or Central Processing Unit, is often referred to as the "brain" of a computer. It is the primary component responsible for interpreting and executing instructions from software programs. The CPU performs basic arithmetic, logic, control, and input/output (I/O) operations that allow a computer to function.

Key Components

- **Control Unit (CU):** Directs the flow of data and instructions between different parts of the computer.
- **Arithmetic Logic Unit (ALU):** Handles mathematical calculations and logical operations.
- **Registers:** Small, high-speed storage locations inside the CPU for quick data access.
- **Cache:** Temporary memory built into the CPU for fast retrieval of frequently used data.

Benefits

- Provides fast and efficient execution of tasks.
- Supports multitasking by handling multiple instructions in sequence or parallel (depending on architecture).
- Acts as the central hub connecting memory, storage, and peripherals.

Applications

- Running operating systems and applications.
- Powering devices like personal computers, smartphones, tablets, and embedded systems.
- Performing computations in AI, data analysis, and gaming.

Summary
The CPU is the central element that drives computer operations. Without it, a computer cannot process instructions or perform tasks. Its speed and efficiency greatly affect the overall performance of any digital device.

CPV (Cost Per View)
Definition: The cost an advertiser pays for each view of a video ad on platforms like YouTube.
Explanation: In my Google Video Ads campaign, I set a $0.05 CPV to promote *IntelliGloss* tutorials, driving traffic to youraistudybuddy.com. Low CPV ensures cost-effective engagement with students and teachers nationwide.
Cross-reference: Google Ads, PPC. See "Promoting Your E-commerce Site."

CTA (Call-to-Action)
Definition: A prompt encouraging users to take action, like "Buy Now."
Explanation: My *IntelliGloss* pages use CTAs linking to `[simpay id="89"]`, driving subscriptions. Clear CTAs in ads and chatbots boost conversions for non-technical audiences.
Cross-reference: Conversion Rate, Payment Form. See "Understanding Stripe."

CTR (Click-Through Rate)
Definition: The percentage of users clicking an ad or link.

Custom Post Types

Definition
Custom Post Types allow users to create content types beyond posts and pages, tailored for specific needs.

Key Components

Post Type registration (`register_post_type`)

Taxonomies (categories/tags)

Admin interface customization

Benefits

Organize content more effectively

Enhances backend management

Makes complex sites easier to navigate

Examples
Creating post types like "Books," "Testimonials," or "Events" for a publishing site or business site.

Applications
Common in real estate, education, portfolios, and directories.

Challenges

Requires some coding knowledge

Not supported by all themes/plugins

Summary
Custom Post Types expand your site's possibilities, helping you structure content to match your goals and user needs.

CUDA (Compute Unified Device Architecture)

Definition
CUDA is a parallel computing platform and programming model developed by NVIDIA. It allows developers to use the power of a Graphics Processing Unit (GPU) not just for graphics, but also for general-purpose computing tasks such as artificial intelligence, scientific simulations, and big data processing.

Key Components

- **Parallel Processing Framework:** Enables thousands of GPU cores to work simultaneously on multiple tasks.
- **CUDA Programming Language:** Extensions of C, C++, and Python that allow developers to write code specifically for GPUs.
- **Runtime & Driver APIs:** Software layers that connect CUDA programs with NVIDIA hardware.
- **Memory Management Tools:** Optimize how data moves between CPU, GPU, and memory.

Benefits

- Speeds up complex calculations by running them in parallel.

- Makes GPUs useful beyond gaming and graphics, especially for AI and machine learning.
- Supported by many AI frameworks like TensorFlow and PyTorch.
- Reduces time for tasks like image recognition, 3D rendering, and simulations.

Applications

- **Artificial Intelligence:** Training and deploying neural networks.
- **Scientific Research:** Simulations in physics, biology, and chemistry.
- **Finance:** Risk modeling and real-time fraud detection.
- **Media:** Video rendering, editing, and streaming.

Summary
CUDA is NVIDIA's key innovation that transformed GPUs into powerful engines for AI, science, and business. By making GPU programming accessible, CUDA has become a cornerstone of modern computing and artificial intelligence.

Custom Post Type

Definition
A Custom Post Type (CPT) is a content type beyond the default WordPress options (post, page) that allows tailored content structure.

Key Components

3. CPT registration (via code or plugin)
4. Custom fields and taxonomies
5. Admin menu integration

Benefits

6. Creates distinct content types (e.g., portfolios, testimonials)
7. Organizes data for specific use cases
8. Enhances flexibility and content modeling

Applications
Common in real estate listings, event calendars, directories, and product catalogs.

Industry
Ecommerce, real estate, media, education

Challenges

9. Requires technical setup or plugin use
10. May break during theme or plugin conflicts

Summary
Custom Post Types extend WordPress beyond blogs and pages, allowing users to organize and display content tailored to their needs.

Customer Lifetime Value (CLV)
The total revenue expected from a customer over their entire relationship with your business, based on their subscription payments.
Example: I calculate CLV for IntelliGloss subscribers on youraistudybuddy.com to estimate long-term profits, helping me plan YouTube content to keep users subscribed. See Stripe: Analytics.

Customer Satisfaction (CSAT)
Definition: A metric measuring how satisfied customers are with your service or product.
Explanation: My AI Study Buddy chatbot improves CSAT on youraistudybuddy.com by answering queries instantly, encouraging *IntelliGloss* purchases. High CSAT (e.g., 90%) boosts retention; I survey users via email to track it.
Cross-reference: AI Chatbot, Conversion Rate. See "Understanding Botpress."

Customizer

Definition
The Customizer is a built-in WordPress tool that allows users to preview and adjust their site's appearance in real time.

Key Components

Theme settings

Site identity (logo, title)

Colors and fonts

Menu and widget management

Benefits

Live preview while editing

No risk of breaking the site

User-friendly interface

Examples
Changing your site title or background color and seeing the changes instantly before publishing.

Applications
Ideal for small business sites, blogs, and personal websites needing quick visual tweaks.

Challenges

Limited by the theme's customization options

May not reflect some plugin features

Summary
The Customizer is your site's control panel for live design edits, enabling safe, real-time adjustments to site appearance.

LETTER D

Dashboard
Definition: A control panel in a CMS or hosting platform for managing website settings.
Explanation: WordPress's dashboard lets me edit pages and add Stripe forms for *IntelliGloss* subscriptions, while Hostinger's hPanel manages domains and backups. Its intuitive layout helps non-technical users avoid navigation issues.
Cross-reference: WordPress, hPanel. See "Understanding WordPress," "Understanding Hostinger."

Database

Definition
The database in WordPress stores all the content, settings, and metadata required to run the site, using MySQL or MariaDB.

Key Components

1. Core tables (wp_posts, wp_users, wp_options, etc.)
2. Structured query language (SQL)
3. Data management tools like phpMyAdmin

Benefits

1. Centralized data storage
2. Efficient content retrieval
3. Supports dynamic websites

Applications
Essential for storing blog posts, product listings, user accounts, and plugin data.

Industry
All industries using dynamic websites

Challenges

1. Vulnerable if not secured
2. Can become bloated over time

Summary
The WordPress database is the digital brain of your site, powering everything from blog posts to login credentials through structured storage.

Digital Marketing
Promoting your business online through strategies like social media, email campaigns, and paid ads to attract customers.
Example: I used digital marketing on YouTube and Facebook to promote *IntelliGloss* and drive traffic to youraistudybuddy.com, boosting sales in 20+ countries. *See Hostinger: SEO.*

Display Network
Definition: Google's network of 2 million+ websites and apps displaying visual ads.
Explanation: I used Display Ads to show *IntelliGloss* banners on education sites, driving awareness to youraistudybuddy.com. These ads target parents and teachers, complementing Search Ads for nationwide reach.
Cross-reference: Google Ads, PPC. See "Promoting Your E-commerce Site."

DNS (Domain Name System)
A system that translates your domain (e.g., youraistudybuddy.com) into an IP address for browsers to find your site.
Example: I updated DNS settings in Hostinger to point youraistudybuddy.com to my *IntelliGloss* store. *See Hostinger: Domains.*

Domain
Definition: The web address of a site, like `youraistudybuddy.com`.
Explanation: I registered my domain via Hostinger's hPanel, ensuring a professional URL for *IntelliGloss* sales. A memorable domain boosts SEO and brand trust for teachers and students searching nationwide.
Cross-reference: Hosting, SEO. See "Understanding Hostinger."

A **domain name** is your website's **address on the internet**.

It's what people type into their browser to visit your website — for example, **youraistudybuddy.com**.

Just like your home has a street address so people can find you, your website needs a domain name so people can find it online.

Some other examples of domain names are:

3. **google.com**
4. **amazon.com**
5. **halasflowers.com** (if you had a site just for your flower shop)

A domain name is unique — **no two websites in the world can have the exact same one**.

You usually buy a domain name from a **domain registrar**, like GoDaddy, Namecheap, or Hostinger (many hosting companies also sell domain names).

LETTER E

E-commerce
Definition: The buying and selling of goods or services online.

Explanation: Youraistudybuddy.com uses WordPress and Stripe to sell *IntelliGloss* subscriptions globally, with my AI Study Buddy guiding users to checkouts. E-commerce enables 24/7 sales for non-technical entrepreneurs.
Cross-reference: Stripe, Subscription Plan. See "Understanding Stripe."

LETTER F

Firewall

Definition: A security system protecting a website from threats.
Explanation: Hostinger's firewall on youraistudybuddy.com safeguards *IntelliGloss* payment pages from hacks. Combined with Stripe's security, it ensures trust for global buyers.
Cross-reference: SSL, Hosting. See "Understanding Hostinger."

Front End
The visible part of a website that users interact with, including design, buttons, and content like product pages or chat windows.
Example: On youraistudybuddy.com, the front end displays *IntelliGloss* product pages and my AI Study Buddy chatbot bubble, making it easy for customers to shop or ask questions. *See WordPress: Appearance, Botpress: Webchat Integration.*

LETTER G

General Intelligence:
The ability to learn, understand, and apply knowledge across a wide variety of tasks and situations, instead of being limited to one specific skill. Humans have general intelligence because we can adapt to new problems, transfer what we know from one area to another, and think creatively in unfamiliar situations. In AI research, "artificial general intelligence" (AGI) refers to machines that could match this human-like flexibility and adaptability.

Google Ads
Definition: Google's advertising platform for creating PPC, Display, and Video Ads.
Explanation: I used Google Ads to promote youraistudybuddy.com, targeting "build a website" searches with Search Ads and *IntelliGloss* with Shopping Ads. It drives nationwide traffic, increasing Starter Plan sign-ups.
Cross-reference: PPC, Display Network, CPV. See "Promoting Your E-commerce Site."

Google Business Profile
Definition: A Google Business Profile is a free online listing from Google that helps your business appear in local search results and on Google Maps. It shows important information like your business name, address, phone number, website, hours, photos, and customer reviews.

A free Google tool to list your business in local search and Maps.
Explanation: My Google Business Profile for Your AI Study Buddy lists youraistudybuddy.com, showcasing *IntelliGloss* to schools nationwide. Adding photos and reviews boosts credibility for

parents and educators.
Cross-reference: SEO, Google Search Console. See "Registering Your Website on Google."

Why It's Important: Having a Google Business Profile increases your visibility when people search for products or services in your area. It also makes your business look more trustworthy and professional.

Example: If you run a flower shop in Los Angeles and someone searches **"flower shop near me"**, your Google Business Profile could appear at the top of the search results with your location, pictures of your flowers, and customer ratings.

Tip: Keep your profile updated with accurate information and fresh photos to attract more customers.

Google Merchant Center
Definition: Google's platform for uploading product data for Shopping Ads.
Explanation: I uploaded *IntelliGloss* details to Merchant Center, enabling Shopping Ads to display its price and image for "educational products" searches. This drives direct sales to youraistudybuddy.com.
Cross-reference: Google Ads, Shopping Ads. See "Promoting Your E-commerce Site."

Google Search Console
Definition: A free tool to monitor and improve your site's visibility in Google search.
Explanation: I used Google Search Console to index youraistudybuddy.com, ensuring pages like "Shop IntelliGloss" appear for "build an e-commerce site." It tracks rankings and fixes indexing issues for nationwide reach.
Cross-reference: SEO, Sitemap. See "Registering Your Website on Google."

GPU (Graphics Processing Unit)

Definition
A GPU, or Graphics Processing Unit, is a specialized processor designed to handle graphics rendering and complex mathematical calculations. While originally built for creating and displaying images, videos, and animations, GPUs are now widely used in artificial intelligence (AI), gaming, data analysis, and other high-performance computing tasks.

Key Components

- **Cores:** Thousands of small processors that work in parallel to handle many tasks at once.
- **Memory (VRAM):** High-speed memory that stores data such as textures, images, and computation results.
- **Processing Units:** Includes pixel shaders, vertex shaders, and tensor cores (in modern GPUs) for different types of calculations.

Benefits

- Speeds up rendering of images, video, and animations.
- Provides parallel computing power for AI training and scientific simulations.
- Enhances performance in gaming, virtual reality, and video editing.

Applications

- **Websites & E-commerce:** Optimizing visuals, 3D product views, and video-based content.
- **AI & Machine Learning:** Training and running neural networks.
- **Creative Work:** Video production, 3D modeling, and design.
- **Everyday Use:** Smooth streaming, fast browsing, and mobile graphics performance.

Summary
The GPU is a powerful companion to the CPU. While the CPU handles general instructions, the GPU excels at parallel processing, making it essential for modern computing tasks that demand speed, graphics, and high performance.

Gutenberg Editor

Definition
The Gutenberg Editor is WordPress's block-based content editor that allows users to design pages and posts using modular blocks instead of traditional text fields.

Key Components

1. Content blocks (paragraphs, images, videos, etc.)
2. Block toolbar
3. Block patterns and reusable blocks
4. Sidebar block settings

Benefits

1. Visual and intuitive content creation
2. Drag-and-drop layout control
3. Easy customization without code

Examples
A blogger can add an image block, followed by a heading and text block, then rearrange them as needed.

Applications
Used by bloggers, marketers, educators, and designers for modern content creation.

Challenges

1. May feel unfamiliar to longtime users of the Classic Editor
2. Some third-party plugins may not be fully compatible

Summary
The Gutenberg Editor revolutionizes how content is built on WordPress, offering flexibility and creativity through a user-friendly block system.

LETTER H

Hosting
Definition: A service that stores your website's files and makes it accessible online.
Explanation: Hostinger hosts youraistudybuddy.com, providing fast servers and hPanel to manage my *IntelliGloss* site. Affordable plans support e-commerce for non-technical users targeting global audiences.
Cross-reference: Domain, hPanel. See "Understanding Hostinger."

hPanel
Definition: Hostinger's control panel for managing hosting, domains, and site settings.
Explanation: I use hPanel to configure youraistudybuddy.com's SSL, backups, and cache, ensuring fast *IntelliGloss* checkouts. Its user-friendly interface simplifies hosting for beginners despite navigation quirks.
Cross-reference: Hosting, WordPress. See "Understanding Hostinger."

LETTER I

Impression: When an ad appears on a user's screen. My *IntelliGloss* Display Ads gained 1,000 impressions daily, boosting awareness. *See Display Network.*

Indexing
Definition: The process of Google adding your site's pages to its search database.
Explanation: I submitted youraistudybuddy.com's sitemap to Google Search Console to index pages like `youraistudybuddy.com/starter-plan-payment`, making them searchable for "educational products." Regular indexing boosts visibility.
Cross-reference: Google Search Console, Sitemap. See "Registering Your Website on Google."

LETTER J

JavaScript
A popular programming language used to make websites interactive. In WordPress, JavaScript powers features like image sliders, pop-up forms, and dynamic content updates without reloading the page.

---

2. JPEG
A common image file format that compresses photos to reduce file size. In WordPress, JPEG images are often used for blog posts, product photos, and banners because they load quickly while keeping decent quality.

---

3. jQuery
A fast, lightweight JavaScript library that simplifies coding for animations, event handling, and AJAX requests. Many WordPress themes and plugins use jQuery to create interactive features without requiring advanced coding.

---

4. JSON (JavaScript Object Notation)
A lightweight data format used to exchange information between a website and a server. In WordPress, JSON is often used in APIs to transfer settings, form data, or content in a structured way.

---

5. Jetpack
A popular WordPress plugin created by Automattic that offers features like security, backups, performance optimization, and site analytics—all in one package.

LETTER K

Keywords
Definition: Keywords are the specific words or short phrases people type into search engines (like Google) when looking for information, products, or services. In WordPress, keywords are used in your website's content, titles, headings, and metadata to help search engines understand what your site is about.

Why They Matter: Choosing the right keywords makes it easier for potential customers to find your site, boosting your visibility and attracting the right audience.

Example: If you have a WordPress site selling handmade candles, some keywords might be **"soy candles," "handmade candles,"** or **"natural scented candles."**

Tip: Use a mix of short keywords (e.g., "candles") and long-tail keywords (e.g., "lavender soy candles for relaxation") for the best SEO results.

Keyword Planner
Definition: Google's tool for researching keywords for ads and SEO.
Explanation: I used Keyword Planner to find "educational products for teachers" for youraistudybuddy.com, optimizing ads and content to rank higher and attract schools nationwide.
Cross-reference: SEO, Google Ads. See "Promoting Your E-commerce Site."

Knowledge Base
Definition: A collection of data a chatbot uses to answer user queries.
Explanation: In Botpress, I trained my AI Study Buddy's knowledge base with youraistudybuddy.com content to answer *IntelliGloss* FAQs, improving customer experience for global users. Regular updates keep responses accurate.
Cross-reference: AI Chatbot, Botpress. See "Understanding Botpress."

LETTER L

Landing Page
A standalone web page designed specifically for a marketing or advertising campaign. Its primary

goal is to guide visitors toward a single call-to-action (CTA), such as signing up for a newsletter, purchasing a product, or downloading a resource.

2. Lazy Loading
A web optimization technique where images, videos, or other media elements load only when they are about to enter the user's viewport. This improves page speed and reduces unnecessary data usage.

3. Link Building
The process of acquiring hyperlinks from other websites to your own. In SEO, quality backlinks from reputable sources can help improve your site's ranking in search engine results.

4. Live Chat Plugin
A WordPress plugin that allows real-time communication between website visitors and support staff. Often used for customer service, sales inquiries, or technical support.

LLM (Large Language Model)

Definition
An LLM, or Large Language Model, is an advanced type of artificial intelligence system designed to understand, generate, and interact with human language. It is trained on massive amounts of text data and uses deep learning techniques to predict the next word in a sequence, enabling it to create coherent sentences, answer questions, summarize information, and even write code.

Key Components

- **Neural Network Architecture:** Usually based on transformers, which excel at handling sequential data like text.
- **Training Data:** Billions of words from books, articles, websites, and other text sources.
- **Parameters:** The "knowledge" of the model stored as weights; large models often have billions or even trillions of parameters.
- **Inference Process:** The method by which the model generates responses or predictions after training.

Benefits

- Provides human-like responses in conversations.
- Can generate original text, translations, summaries, and explanations.
- Supports a wide range of industries including education, healthcare, customer service, and research.

Applications

- **Chatbots & Virtual Assistants:** Powering tools like ChatGPT and other AI assistants.
- **Content Creation:** Writing articles, blogs, or marketing copy.
- **Programming Help:** Generating or debugging code.
- **Data Analysis:** Summarizing large volumes of information.

Summary
An LLM is the backbone of modern conversational AI. By learning patterns from enormous text datasets, it can understand context, respond intelligently, and perform tasks once thought unique to humans.

5. LSI Keywords (Latent Semantic Indexing Keywords)
Words or phrases that are conceptually related to the main keyword. Using LSI keywords in content can help search engines better understand the topic and improve SEO performance.

LETTER M

Media Library

Definition
The Media Library is the built-in WordPress tool for uploading, managing, and inserting images, videos, and documents.

Key Components

Upload interface

Attachment details panel

Media filters and search

Benefits

Central storage of all media assets

Easy to reuse images/files

Drag-and-drop upload

Examples
A user uploads a photo once and inserts it into multiple blog posts without re-uploading.

Applications
Essential in blog, portfolio, and eCommerce sites to manage visual content.

Challenges

Can become disorganized without folders

Limited file editing tools

Summary
The Media Library keeps all your files organized and accessible, making it simple to manage and reuse digital assets.

Menus

Definition
Menus in WordPress provide navigational links that allow users to explore different pages of your site.

Key Components

Menu editor interface

Drag-and-drop items

Custom links and nested menus

Benefits

Creates structured navigation

Improves user experience

SEO-friendly navigation

Examples
A site may have a main menu (Home, About, Services, Contact) and a footer menu (Privacy Policy, Terms, Sitemap).

Applications
Every WordPress site uses menus to guide visitors to key content.

Challenges

Overcrowded menus confuse users

Improper hierarchy can hurt SEO

Summary
Menus are the roadmaps of your site—helping visitors reach content easily and keeping your site's structure organized.

Menu Editor

Definition
The Menu Editor in WordPress allows users to create, edit, and organize site navigation menus for pages, posts, categories, or custom links.

Key Components

1. Drag-and-drop interface
2. Menu location settings
3. Link management and nesting options

Benefits

1. Enhances site usability
2. Easy to modify and update
3. Organizes content flow

Applications
Used in all websites to structure user navigation and guide visitors to key pages.

Industry
All industries with digital presence

Challenges

1. Overloaded menus can confuse users
2. Some themes limit menu flexibility

Summary
Menu Editor helps users build and manage clear and functional site navigation, improving the user experience and site structure.

Meta Tag
Definition: HTML code describing a page's content for search engines.
Explanation: I used Yoast SEO to add meta tags like "Build E-commerce with IntelliGloss" to youraistudybuddy.com, improving rankings for "build a website." Meta tags guide Google to relevant pages.
Cross-reference: SEO, Google Search Console. See "Registering Your Website on Google."

MRR (Monthly Recurring Revenue)
The predictable monthly income generated from active subscriptions, helping you forecast revenue for your e-commerce business.
Example: My IntelliGloss subscription plans on youraistudybuddy.com generate MRR, tracked via Stripe, supporting my global expansion to 20+ countries. See Stripe: Analytics.

MRR (Monthly Recurring Revenue)
The predictable monthly income generated from active subscriptions, helping you forecast revenue for your e-commerce business.
Example: My IntelliGloss subscription plans on youraistudybuddy.com generate MRR, tracked via Stripe, supporting my global expansion to 20+ countries. See Stripe: Analytics

Multisite Network

Definition
Multisite is a WordPress feature that allows you to run multiple websites from a single WordPress installation.

Key Components

Network Admin Dashboard

Site creation and management tools

Shared plugins and themes

Benefits

Centralized updates and management

Resource efficiency

Scalable for large organizations

Examples
A university can manage separate sites for each department from one central system.

Applications
Common in educational institutions, corporations, and large blog networks.

Challenges

Requires advanced configuration

Plugin compatibility may vary

Summary
Multisite streamlines the management of multiple WordPress sites, making it an ideal solution for networks and large organizations.

LETTER N

NLP (Natural Language Processing)
Definition: AI technology enabling chatbots to understand human language.
Explanation: Botpress's NLP powers my AI Study Buddy to interpret queries like "What's IntelliGloss?" and respond naturally, guiding users to payment pages. It supports 90+ languages for global reach.
Cross-reference: AI Chatbot, Botpress. See "Understanding Botpress."

Neural Network:
A type of computer system modeled after the way the human brain works. It is made up of layers

of small "nodes" (like artificial neurons) that pass information to each other. Each node takes in data, processes it, and sends the result to the next layer. By adjusting the strength of the connections between nodes (called *weights*), a neural network can learn patterns — such as recognizing faces, translating languages, or predicting outcomes.

The Neural Network: The Digital Brain

A **neural network** is the engine that powers most modern AI. It's inspired by how the human brain works, but in a much simpler way. Instead of billions of brain cells, a digital neural network is made up of tiny units called **neurons**.

How It Works

- **Inputs (data):** Numbers come in — for example, pixels from an image or words from a sentence.
- **Weights (importance):** Each input is multiplied by a weight that tells the network how important that piece of information is.
- **Layers:** Neurons are stacked in layers. The input layer takes in raw data, hidden layers do the learning, and the output layer gives the answer.
- **Output (result):** After passing through the layers, the network produces a final decision — "cat" or "dog," "yes" or "no," "word A" or "word B."

How It Learns

At first, the network guesses randomly. It compares its answer to the correct one, sees the error, and adjusts its weights (a process called **backpropagation**). Over many rounds, it gets better and better — just like a person learning from practice.

A Simple Example

Imagine teaching a child the difference between a cat and a dog:

- At first, the child guesses randomly.
- Each time you correct them, they pay attention to different details — whiskers, ears, tails, sounds.
- After many examples, the child learns the pattern.

That's exactly how a neural network improves with training.

🌸 **Plain English:**
A neural network is like a digital brain. It learns patterns from examples and gets better the more data you give it. That's why AI can recognize faces, translate languages, and power chatbots — all thanks to neural networks.

How Neural Networks Connect to AI

Neural networks are the **core technology that makes modern AI possible.** Without them, AI would just be a set of fixed rules written by humans. With them, AI can learn, adapt, and improve on its own.

The Connection

- **AI is the big picture.** It means machines doing tasks that normally require human intelligence (like talking, recognizing images, or making decisions).
- **Neural networks are the brain inside AI.** They are the method that allows AI to actually learn patterns from data.

Examples in Action

- A **chatbot** uses a neural network to understand your words and reply naturally.
- **Image recognition AI** uses neural networks to detect faces, objects, and handwriting.
- **Translation AI** uses neural networks to convert one language into another.
- **Recommendation systems** (like Netflix or YouTube) use them to suggest what you'll like next.

Why It Matters

Before neural networks, AI was limited — it could only do what programmers told it. Now, with neural networks, AI can *teach itself from data,* which is why AI feels so much smarter today.

🌸 **Plain English:**
AI is the whole system. Neural networks are the digital brain that make it work.

How a Neural Network Gets Data

A neural network doesn't magically create knowledge — it **learns from data we give it.** The data usually comes from three main sources:

1. Prepared Datasets (Training Data)

- Researchers or developers collect big sets of examples.
- Example: thousands of cat and dog photos, or millions of sentences in English and French.
- The network studies these examples over and over until it learns the patterns.

2. Live Inputs (User Data)

- After training, the network can take in **new data from users**.
- Example: when you type a question into a chatbot, your words become the input data.
- The network processes this input and produces an output (the answer).

3. Feedback & Corrections

- Sometimes, neural networks get feedback to improve.
- Example: if a system predicts wrong, the correction is stored and helps it adjust its "weights."
- This is how the system keeps learning.

🏵 Plain English

Neural networks get their data from **what people feed them** — large collections of examples for training, and real-time inputs (like your questions or images) when they're running.

They don't "think" on their own — they need lots of examples to learn from.

LETTER O

Open Source
Software with publicly accessible source code, allowing anyone to view, modify, and distribute it. WordPress is open-source, meaning developers worldwide can contribute to its improvement and create plugins, themes, and custom features.

Optimization
The process of improving a website's performance, speed, and visibility. In WordPress, optimization includes compressing images, minimizing code, and using SEO techniques to boost search rankings.

Outbound Link
A hyperlink that directs visitors from your website to another external website. Outbound links can provide additional resources for readers and may improve SEO when used strategically.

Overlay
A visual layer that appears on top of website content, often used for pop-ups, lightboxes, or promotional banners in WordPress themes and plugins.

5. Organic Traffic
Website visitors who find your site naturally through search engines, rather than via paid advertisements. Increasing organic traffic is a major goal of SEO in WordPress site management.

LETTER P

Page Builder

Definition
A Page Builder is a plugin or theme feature that enables drag-and-drop design of web pages without coding.

Key Components

1. Visual design interface
2. Pre-designed elements (text, images, buttons)
3. Section and column layouts
4. Mobile responsiveness controls

Benefits

1. No coding required
2. Real-time visual editing
3. Speeds up the design process

Examples
Popular page builders include Elementor, Beaver Builder, and WPBakery.

Applications
Used to build landing pages, homepages, and custom layouts quickly and easily.

Challenges

1. Can slow down the site if overused
2. Switching builders later may cause layout issues

Summary
Page Builders empower users to visually construct beautiful pages with ease, giving creative control to non-developers.

Payment Gateway
A service that securely processes online payments, like credit cards or digital wallets, connecting your store to banks.
Example: I used Stripe as my payment gateway on youraistudybuddy.com to process *IntelliGloss* purchases safely and support global customers. *See Stripe: Connecting to Your Website.*

Payment Form

Definition: A web form for collecting payments, often embedded via shortcode.
Explanation: I embedded WP Simple Pay's `[simpay id="89"]` on youraistudybuddy.com for *IntelliGloss* Starter Plan payments, streamlining checkouts. Testing forms prevents redirect errors, a lesson from my setup struggles.
Cross-reference: Stripe, Subscription Plan. See "Understanding Stripe."

Payout

Funds transferred from your payment processor to your bank account after sales.
Example: Stripe deposits *IntelliGloss* sales to my bank weekly, supporting my YouTube channel. *See Stripe: Linking Your Bank Account.*

Permalink

Definition
A permalink (permanent link) is the full URL of a WordPress page, post, or archive that is used for linking and indexing.

Key Components

1. URL structure
2. Slug (editable part of the URL)

3. Permalink settings panel

Benefits

1. SEO-friendly links
2. Easier to remember and share
3. Better indexing by search engines

Applications
Used in blogs, eCommerce sites, and educational platforms to organize content.

Industry
SEO, journalism, blogging, retail

Challenges

1. Changing permalinks can break links
2. Poorly structured URLs can hurt SEO

Summary
Permalinks are essential for web visibility and structure, serving as the digital address of every piece of content on a WordPress site.

PPC (Pay-Per-Click)

Definition: An ad model where you pay each time a user clicks your ad.
Explanation: I used Google Ads PPC to promote youraistudybuddy.com, targeting "build a chatbot" searches to drive *IntelliGloss* sales. A $2–$5 cost-per-click delivers high-intent traffic nationwide.
Cross-reference: Google Ads, Quality Score. See "Promoting Your E-commerce Site."

Plugin

Definition
A plugin is a piece of software that adds specific features and functionality to a WordPress site without altering core files. A software add-on that extends WordPress functionality.
Explanation: On youraistudybuddy.com, I use WP Simple Pay and Botpress plugins to add payment forms and my AI Study Buddy chatbot. Plugins simplify e-commerce for non-technical users but require updates to avoid glitches.
Cross-reference: WordPress, Botpress. See "Understanding WordPress."

Key Components

1. PHP scripts
2. Plugin dashboard
3. Activation and update systems
4. Hooks and filters

Benefits

1. Enhances website capabilities
2. Adds features like SEO tools, contact forms, and backups
3. Saves development time

Applications
Common in business, blogging, nonprofit, and educational websites.

Industry
Digital marketing, development, eLearning, service-based industries

Challenges

1. Too many plugins can slow down the site
2. Conflicts between plugins

Summary
Plugins are essential for extending WordPress capabilities and tailoring the site to meet specific user or business needs.

Permalink: A permanent URL for a webpage. I fixed permalinks to avoid redirects on *IntelliGloss* pages. *See Redirect.*

Platform
A software foundation for building and managing a website, like WordPress or Shopify, that supports e-commerce features.
Example: I chose WordPress as my platform for youraistudybuddy.com to integrate WooCommerce for *IntelliGloss* sales and Botpress for my chatbot. *See WordPress: Dashboard.*

Plugins

Definition
Plugins are add-on tools that extend the functionality of a WordPress site without altering the core code.

Key Components

1. Plugin Repository
2. Installed Plugins tab
3. Activation/Deactivation options
4. Plugin settings interface

Benefits

1. Adds custom features (SEO, security, eCommerce, etc.)
2. Saves time and reduces need for coding
3. Easy to install and manage

Examples
WooCommerce (eCommerce), Yoast SEO (search engine optimization), WPForms (form builder).

Applications
Used for everything from performance optimization to marketing automation and design customization.

Challenges

1. Too many plugins can slow down the site
2. Conflicts between plugins can cause site errors

Summary
WordPress plugins are powerful extensions that transform simple sites into functional, tailored platforms with specialized capabilities.

LETTER Q

Quality Score

Definition: Google's rating of an ad's relevance and landing page quality.
Explanation: A high Quality Score for my *IntelliGloss* ads lowered costs in Google Ads, as my landing page matched "educational products" keywords. Optimizing content and links boosts scores.
Cross-reference: PPC, Google Ads. See "Promoting Your E-commerce Site."

LETTER R

RAM

RAM (Random Access Memory) is a type of computer memory that stores data temporarily while the computer is running. It gives your device a fast workspace to load apps, open files, and run programs. When the power is turned off, everything in RAM is erased.

Key Points

- **Temporary storage:** Keeps data only while the computer is on.
- **Fast access:** Much faster than hard drives or SSDs.
- **Purpose:** Lets the CPU quickly use data it needs right now.
- **Volatile:** Data disappears when the computer shuts down.

Plain English

RAM is your computer's **short-term memory** — it helps your device work quickly by holding the stuff you're using right now.

Redirect

Definition: A setting that sends users from one URL to another.
Explanation: I fixed a redirect error on youraistudybuddy.com when my Starter Plan page misdirected, using Hostinger's cache clear and WordPress permalinks. Correct redirects ensure seamless navigation to `[simpay id="89"]`.
Cross-reference: SEO, Google Search Console. See "Registering Your Website on Google."

Responsive Design

Definition: A website layout that adapts to different devices.
Explanation: Youraistudybuddy.com's WordPress theme ensures *IntelliGloss* pages look great on phones, boosting SEO and user experience for students browsing on mobile.
Cross-reference: SEO, WordPress. See "Understanding WordPress."

REST API

Definition
The WordPress REST API allows developers to interact with site data using HTTP requests, enabling external applications to read and write content.

Key Components

1. Endpoints for posts, pages, users, etc.
2. JSON data structure
3. Authentication and permissions

Benefits

1. Enables mobile app and third-party integrations
2. Separates front-end and back-end
3. Supports headless CMS design

Applications
Used in mobile app development, custom dashboards, and cross-platform publishing.

Industry
App development, enterprise CMS, digital publishing

Challenges

1. Requires technical expertise
2. Security risks if not implemented properly

Summary
The REST API turns WordPress into a flexible data source, unlocking possibilities for app integration and modern web architecture.

Retargeting: Ads targeting users who visited your site. I retargeted *IntelliGloss* visitors with Display Ads, recovering 10% of abandoned carts. *See Google Ads.*

Robots.txt: A file controlling which pages search engines crawl. I edited robots.txt in Hostinger to index youraistudybuddy.com fully. *See Indexing.*

Roles and Permissions

Definition
WordPress roles and permissions determine what each user can and cannot do on the site.

Key Components

1. Default roles: Administrator, Editor, Author, Contributor, Subscriber
2. Capabilities: read, edit, delete, publish, manage

Benefits

1. Improved security through access control
2. Encourages team collaboration
3. Prevents accidental changes to critical settings

Examples
An Editor can publish posts, but a Contributor can only submit drafts.

Applications
Used in news sites, blogs with guest authors, and membership sites.

Challenges

1. Custom roles require technical setup
2. Misassigning roles can lead to errors or vulnerabilities

Summary
WordPress roles ensure that every user has the right level of access, keeping your site secure and organized.

Recurring Billing
Definition: Automatic, regular payments for subscriptions.
Explanation: Stripe's recurring billing powers my $29.95/month Family Premium plan on youraistudybuddy.com, ensuring steady revenue. I set 7-day trials to attract users, reducing churn for *IntelliGloss*.
Cross-reference: Subscription Plan, MRR. See "Understanding Stripe."

Roles and Permissions

Definition
WordPress roles and permissions determine what each user can and cannot do on the site.

Key Components

3. Default roles: Administrator, Editor, Author, Contributor, Subscriber
4. Capabilities: read, edit, delete, publish, manage

Benefits

5. Improved security through access control

6. Encourages team collaboration
7. Prevents accidental changes to critical settings

Examples
An Editor can publish posts, but a Contributor can only submit drafts.

Applications
Used in news sites, blogs with guest authors, and membership sites.

Challenges

8. Custom roles require technical setup
9. Misassigning roles can lead to errors or vulnerabilities

Summary
WordPress roles ensure that every user has the right level of access, keeping your site secure and organized.

LETTER S

SEO (Search Engine Optimization)
Techniques to improve your website's ranking on search engines like Google to attract more visitors.
Example: I optimized youraistudybuddy.com with keywords like "AI dictionary" to drive traffic to *IntelliGloss* and my YouTube videos. *See Hostinger: SEO.*

SSL Certificate
A digital certificate that secures your site with "https://" and encrypts data.
Example: Hostinger's free SSL ensured secure *IntelliGloss* transactions on youraistudybuddy.com. *See Hostinger: Security.*

Subscription Model
A business model where customers pay a recurring fee (e.g., monthly) for access to products or services, supported by AI agents like chatbots to drive sign-ups and engagement.
Example: On youraistudybuddy.com, my AI Study Buddy chatbot promotes IntelliGloss subscriptions by answering questions and guiding users to Stripe checkouts, boosting sales in 20+ countries. See Botpress: Webchat Integration, Stripe: Connecting to Your Website.

Subscription Tier
Different levels of subscription plans offering varied features or benefits, allowing customers to choose based on their needs.
Example: On youraistudybuddy.com, I offer basic and premium IntelliGloss subscription tiers, with my AI Study Buddy guiding users to select the right plan. See Botpress: Webchat Integration, Stripe: Connecting to Your Website.

Schema Markup: Code enhancing search result displays. I added schema to *IntelliGloss* pages for richer Google snippets. *See SEO.*

SEO (Search Engine Optimization)
Definition: Techniques to improve a site's ranking in search results.
Explanation: I optimized youraistudybuddy.com with keywords like "build an e-commerce site" using Yoast SEO, boosting organic traffic for *IntelliGloss*. SEO drives long-term visibility for teachers and students.
Cross-reference: Google Search Console, Meta Tag. See "Registering Your Website on Google."

Shopping Ads
Definition: Google Ads displaying product images, prices, and links.
Explanation: My *IntelliGloss* Shopping Ads in Google Merchant Center show its $29.95 price, driving direct sales on youraistudybuddy.com. They target "educational products" searches for nationwide reach.
Cross-reference: Google Merchant Center, Google Ads. See "Promoting Your E-commerce Site."

Shortcode
Definition: A small code snippet in WordPress to embed features like payment forms.
Explanation: I used `[simpay id="89"]` to embed Stripe payment forms for *IntelliGloss* subscriptions on youraistudybuddy.com. Shortcodes simplify adding complex features for non-technical users.
Cross-reference: Payment Form, Stripe. See "Understanding Stripe."

Site Health

Definition
Site Health is a WordPress feature that checks for common performance and security issues, offering recommendations for improvement.

Key Components

1. Health status overview
2. Recommended improvements
3. Debug information

Benefits

1. Monitors performance and security
2. Identifies configuration issues
3. Helps site maintenance

Applications
Useful for website administrators to proactively maintain site health.

Industry
All industries, especially IT, ecommerce, and education

Challenges

1. Advanced issues may be hard to understand
2. False positives from custom setups

Summary
Site Health provides diagnostic tools to maintain and optimize a WordPress site's performance and security.

Sitemap
Definition: A file listing a website's pages for search engines to crawl.
Explanation: Yoast SEO generated a sitemap for youraistudybuddy.com, submitted to Google Search Console, ensuring *IntelliGloss* pages rank for "build a website." Regular updates keep all pages indexed.
Cross-reference: Google Search Console, Indexing. See "Registering Your Website on Google."

Subscription Plan
Definition: A recurring payment model for ongoing services or products.
Explanation: My $9.95/month Starter Plan on youraistudybuddy.com, powered by Stripe, offers *IntelliGloss* access with a 7-day trial. Clear subscription pages drive conversions for teachers and parents.
Cross-reference: Recurring Billing, Payment Form. See "Understanding Stripe."

Shortcodes

Definition
Shortcodes are small code snippets enclosed in brackets that perform dynamic functions within WordPress content.

Key Components

1. Shortcode API
2. Built-in and custom shortcodes
3. Shortcode generators in plugins

Benefits

1. Add complex features easily
2. Keeps post/page content clean
3. Saves time and improves efficiency

Examples
`[gallery]` for image displays, `[video]` to embed videos, or `[contact-form-7]` for contact forms.

Applications
Used in blog posts, landing pages, and product descriptions to insert elements like sliders, forms, or testimonials.

Challenges

1. Not beginner-friendly without guidance
2. Plugin shortcodes can break if plugin is deactivated

Summary
Shortcodes are time-saving tools that pack powerful features into a compact format, ideal for adding rich content with minimal effort.

Staging Environment

Definition
A staging environment is a private clone of your live website used to test changes before making them public.

Key Components

1. Cloned site copy
2. Password-protected access
3. Sync options between staging and live site

Benefits

1. Safe testing zone
2. Prevents live site disruptions
3. Ideal for plugin/theme updates and redesigns

Examples
Testing a new theme on a staging site before pushing it live.

Applications
Used by developers, designers, and site owners during updates or redesign phases.

Challenges

1. Requires extra hosting resources
2. Syncing issues may arise if changes aren't tracked properly

Summary
Staging environments are the safety nets of website development, allowing you to experiment without putting your live site at risk

Superintelligence:
A form of intelligence that vastly surpasses the best human brains in nearly every field, including creativity, problem-solving, scientific discovery, and social skills. It is a theoretical stage of AI development where machines would not only match but greatly exceed human intelligence. Superintelligence is often discussed in philosophy, ethics, and future studies as both a potential breakthrough and a risk, since its goals and actions could be beyond human control.

Synthetic Intelligence

Synthetic Intelligence is another term for **artificial intelligence (AI)** — intelligence created by humans rather than occurring naturally. It refers to computer systems or machines that are designed to mimic human thought, reasoning, and problem-solving.

Key Points

- **"Synthetic" means man-made.** Just like synthetic fabric is created in a factory, synthetic intelligence is built by humans.
- It includes all kinds of AI: from simple chatbots to advanced neural networks.
- Some people use the term to highlight the difference between **biological intelligence** (humans, animals) and **machine intelligence** (AI).

🏵 Plain English

Synthetic intelligence simply means **human-made intelligence** — machines that can think or learn in ways similar to us.

LETTER T

Taxonomy

Definition
Taxonomy in WordPress refers to the classification of content using categories, tags, or custom groupings.

Key Components

1. Categories
2. Tags
3. Custom taxonomies
4. Hierarchical structure

Benefits

1. Improves site organization
2. Enhances SEO
3. Facilitates user navigation

Applications
Used for organizing blog posts, product listings, and media files.

Industry
Retail, content marketing, education

Challenges

1. Misused tags can reduce clarity
2. Redundant taxonomies cause confusion

Summary
Taxonomies help structure content and improve discoverability, making them essential for large or information-rich websites.

Theme

Definition: A template defining a website's design and layout.
Explanation: I used the Astra theme on youraistudybuddy.com for a clean *IntelliGloss* storefront, customizable without coding. Themes enhance user experience for non-technical builders.
Cross-reference: WordPress, CMS. See "Understanding WordPress."

TLD (Top-Level Domain)
The last part of a domain name, like .com or .shop, indicating its purpose.
Example: I chose .com for youraistudybuddy.com to signal global credibility for *IntelliGloss. See Domain Name Companies.*

LETTER U

User Intent: The goal behind a user's search. My chatbot aligns with user intent for "build a chatbot," driving conversions. *See AI Chatbot.*

Here are 10 original WordPress glossary entries, each with detailed sections including definition, key components, benefits, examples, challenges, and a summary. Let me know if you'd like these exported to a PDF, added to your book, or formatted for the front family section.

User Roles

Definition
User Roles in WordPress define what actions each user can perform, controlling access and responsibilities on the website.

Key Components

1. Default roles: Administrator, Editor, Author, Contributor, Subscriber
2. Role-specific capabilities
3. Role management via dashboard or plugins

Benefits

1. Enhances site security
2. Delegates responsibilities
3. Streamlines collaboration

Applications
Used in team blogs, educational sites, company intranets, and client-based portals.

Industry
Media, education, corporate websites

Challenges

1. Misassigned roles may lead to security risks
2. Limited flexibility in default roles

Summary
User roles allow for controlled access and responsibilities within a WordPress site, making multi-user management secure and efficient.

LETTER V

Visual Editor
The WordPress content editing interface that lets you create and format posts and pages without writing code. Known as the Block Editor (Gutenberg) in modern WordPress, it uses blocks for text, images, videos, and other content elements.

Version Control
A system that tracks changes made to your website's files, themes, and plugins. It allows you to revert to previous versions if something goes wrong. Git is a popular tool for version control in WordPress development.

Visitor Analytics
Data that tracks how users interact with your site, including page views, session duration, bounce rates, and geographic location. WordPress site owners often use tools like Google Analytics to gather this information.

Validation
The process of checking your website's HTML, CSS, and other code against web standards to ensure it displays correctly across browsers and devices. Validation helps maintain a professional and accessible WordPress site.

Virtual Private Server (VPS)
A type of hosting where a physical server is divided into virtual compartments, each with its own resources. VPS hosting offers more control and performance for WordPress sites than shared hosting.

LETTER W

Webchat
A chatbot interface embedded on your website, appearing as a chat bubble.
Example: My AI Study Buddy webchat answers queries and links to *IntelliGloss* on youraistudybuddy.com. *See Botpress: Webchat Integration.*

Webhook
Definition: A tool that sends data between apps when an event occurs.
Explanation: I used Stripe webhooks to trigger AI Study Buddy messages after *IntelliGloss* payments, enhancing user experience on youraistudybuddy.com. Webhooks automate tasks without coding.
Cross-reference: AI Chatbot, Stripe. See "Understanding Botpress," "Understanding Stripe."

Web Hosting

Definition
Web hosting refers to the service of storing your WordPress website's files on a server, making it accessible on the internet.

Key Components

1. Server space
2. Bandwidth
3. Security features
4. Technical support

Benefits

1. Makes your site live and accessible
2. Provides storage and speed
3. Offers data backups and uptime monitoring

Applications
Used by all types of WordPress websites—from blogs to online stores.

Industry
All industries, especially technology, commerce, and education

Challenges

1. Downtime risks with poor hosting
2. Limited resources on shared plans

Summary
Web hosting is the digital foundation of any WordPress site, providing the performance and support needed for stable online presence.

Website
An online collection of pages accessible via a domain, used for sharing content or selling products.
Example: Youraistudybuddy.com is my website for selling *IntelliGloss* and hosting my AI Study Buddy chatbot, driving global engagement. *See WordPress: Pages.*

WHOIS Privacy
A service that hides your contact info from public domain records.
Example: I used Namecheap's WHOIS privacy to protect my details for youraistudybuddy.com.
See Domain Name Companies.

Widget
Definition: A small WordPress feature, like a chatbot or menu, added to a site.
Explanation: I added my AI Study Buddy widget to youraistudybuddy.com's sidebar, guiding users to subscriptions. Widgets simplify functionality for non-technical users.
Cross-reference: Plugin, Botpress. See "Understanding WordPress."

WooCommerce
A WordPress plugin that adds e-commerce features like product pages and checkouts.
Example: I used WooCommerce to sell *IntelliGloss* and manage orders on youraistudybuddy.com. *See WordPress: WooCommerce.*

WordPress
A platform for building websites, customizable with plugins for e-commerce and chatbots.
Example: I built youraistudybuddy.com on WordPress to host my bot and *IntelliGloss* store. *See WordPress: Dashboard.*

WordPress Dashboard

Definition:
The dashboard is the central admin interface where users manage all aspects of their WordPress site.

Key Components:

Left sidebar with menu items

Site health and status widgets

Access to posts, pages, themes, plugins, and settings

Quick draft and activity panels

Benefits:

Simplifies content creation and updates

Intuitive for beginners

Allows full control of your website

Customizable widgets for quick access

Examples:
You can use the dashboard to publish blog posts, install plugins, and change site settings—all in one place.

Challenges:

Can be overwhelming with too many plugins

Might slow down if hosting is poor

Requires learning curve for new users

Limited mobile optimization

Summary:
The WordPress dashboard is your command center, giving you full control over design, content, and functionality without needing to code.

WordPress Gutenberg Editor

Definition:
The Gutenberg editor (also known as the block editor) allows users to build content using drag-and-drop blocks instead of one large text area.

Key Components:

Content blocks (paragraph, image, list, etc.)

Reusable blocks

Drag-and-drop design

Full-site editing (FSE) support

Benefits:

Beginner-friendly

Flexible layout control

Faster content formatting

Visual preview of content

Examples:
You can create a landing page with headers, images, and buttons without using HTML.

Challenges:

Not all plugins/themes support it

Some users prefer the Classic Editor

Can be buggy with third-party blocks

Learning curve for advanced formatting

Summary:
Gutenberg gives users more design freedom with a visual interface, making it easier to create stylish, engaging content.

WordPress Media Library

Definition:

The Media Library is a central place to upload, store, and manage all image, video, audio, and document files on your site.

Key Components:

1. Upload button
2. File sorting and filtering
3. Image editing tools
4. Attachment details

Benefits:

1. Keeps media organized
2. Allows reuse of files
3. Reduces need for external hosting
4. Simple interface

Examples:
You upload your logo once and insert it into multiple pages and posts.

Challenges:

1. Limited file types unless adjusted
2. Large files can slow down site speed
3. Needs regular cleanup to prevent clutter

Summary:
The Media Library makes it easy to manage your visuals and documents, creating a seamless content workflow.

WordPress Menu Navigation

Definition:
Menus in WordPress are custom navigation links that organize pages, posts, categories, and more for users.

Key Components:

1. Menu creation panel
2. Drag-and-drop menu builder
3. Dropdown and nested items
4. Theme location assignment

Benefits:

1. Enhances user experience
2. Highlights important content
3. Flexible menu locations
4. Supports custom links

Examples:
A top menu with "Home," "About," "Shop," and "Contact" links helps visitors explore your site.

Challenges:

1. Poorly structured menus confuse users
2. Limited design if theme doesn't support customization
3. Complex sites may need mega menus

Summary:
Menus guide your visitors. Keeping them simple and well-structured makes it easy for users to navigate your content

WordPress Theme

Definition:
A WordPress theme is a pre-designed template that controls the appearance and layout of a WordPress website.

Key Components:

Template files (PHP)

CSS stylesheets

Theme customizer options

Functions.php file

JavaScript for interactivity

Benefits:

Offers professional-looking design without coding

Customizable to match your brand

Responsive and mobile-friendly options available

Speeds up development time

Examples:
A wedding blog might use the "Astra" theme for elegance and minimalism. An ecommerce store might choose "OceanWP" for WooCommerce integration.

Challenges:

Some themes are bloated and slow down your site

Compatibility issues may arise with plugins

Premium themes can be costly

Limited flexibility with rigid templates

Summary:
Themes are the visual backbone of a WordPress site. Choosing a well-coded, lightweight theme can enhance user experience and SEO while maintaining aesthetic appeal.

WordPress Page vs. Post

Definition:
Posts are time-stamped content pieces (like blog entries), while pages are static content like "About Us" or "Contact."

Key Components:

1. Posts appear in reverse chronological order
2. Pages are hierarchical and timeless
3. Both use the WordPress editor
4. Can include images, media, and links

Benefits:

1. Posts are great for ongoing content
2. Pages help structure a site
3. Both improve site organization
4. Pages can be used for landing or product info

Examples:
A blog article is a post; a FAQ section is a page.

Challenges:

1. Confusion between when to use each
2. Posts require regular updates
3. Too many pages can overwhelm navigation
4. Posts can become outdated quickly

Summary:
Knowing the difference between posts and pages helps create a well-organized site that's both dynamic and structured.

WordPress Permalinks

Definition:
Permalinks are the permanent URLs of your posts and pages, controlling how site links are displayed.

Key Components:

1. Structure settings (e.g., /blog/%postname%/)
2. SEO-friendly URLs
3. .htaccess file (for Apache servers)
4. Slug customization

Benefits:

1. Improves search engine visibility
2. Clean and readable URLs
3. Easy sharing of links
4. Helps with site structure

Examples:
`yoursite.com/2025/ai-dictionary-release/` is cleaner than `yoursite.com/?p=123`.

Challenges:

1. Changing permalink structure can break links
2. Requires technical knowledge to fix redirection issues
3. Poor structure can confuse users and search engines

Summary:
Permalinks are essential for SEO and user navigation. Set them up properly at the start to avoid future headaches.

WordPress Plugin

Definition:
A plugin is a software add-on that extends the core functionality of a WordPress site.

Key Components:

1. Plugin files in PHP
2. Admin interface for configuration
3. Hooks and filters
4. Plugin activation and updates system

Benefits:

1. Adds features like SEO tools, e-commerce, contact forms
2. Reduces the need for custom coding
3. Modular—use only what you need
4. Many free options available

Examples:
"Yoast SEO" improves search engine visibility. "WooCommerce" turns your site into an online store.

Challenges:

1. Too many plugins can slow down your site
2. Some plugins may conflict with each other
3. Security vulnerabilities if not updated
4. May require technical support

Summary:
Plugins turn a basic WordPress site into a powerful platform tailored to your needs, from blogs to full-fledged online businesses.

WordPress Theme
Definition:
A WordPress theme is a pre-designed template that controls the appearance and layout of a WordPress website.

Key Components:

Template files (PHP)

CSS stylesheets

Theme customizer options

Functions.php file

JavaScript for interactivity

Benefits:

Offers professional-looking design without coding

Customizable to match your brand

Responsive and mobile-friendly options available

Speeds up development time

Examples:
A wedding blog might use the "Astra" theme for elegance and minimalism. An ecommerce store might choose "OceanWP" for WooCommerce integration.

Challenges:

Some themes are bloated and slow down your site

Compatibility issues may arise with plugins

Premium themes can be costly

Limited flexibility with rigid templates

Summary:
Themes are the visual backbone of a WordPress site. Choosing a well-coded, lightweight theme can enhance user experience and SEO while maintaining aesthetic appeal.

WordPress Widget

Definition:
A widget is a small content block that can be added to sidebars, footers, and other widget-ready areas.

Key Components:

1. Drag-and-drop interface
2. Widget-ready theme areas
3. Content elements like menus, text, or calendars
4. Plugin-generated widgets

Benefits:

1. Easy to add features like recent posts or contact forms
2. Enhances site interactivity
3. No coding needed
4. Custom placement options

Examples:
A footer widget might include your business contact info. A sidebar widget could show your latest blog posts.

Challenges:

1. Not all themes support widget areas
2. Can become cluttered if overused
3. Limited layout control
4. Widgets don't always display consistently on mobile

Summary:
Widgets offer quick customization for non-tech users, helping enhance user engagement and design without coding knowledge.

WordPress Page vs. Post

Definition:
Posts are time-stamped content pieces (like blog entries), while pages are static content like "About Us" or "Contact."

Key Components:

5. Posts appear in reverse chronological order
6. Pages are hierarchical and timeless
7. Both use the WordPress editor
8. Can include images, media, and links

Benefits:

9. Posts are great for ongoing content
10. Pages help structure a site
11. Both improve site organization
12. Pages can be used for landing or product info

Examples:
A blog article is a post; a FAQ section is a page.

Challenges:

13. Confusion between when to use each
14. Posts require regular updates
15. Too many pages can overwhelm navigation
16. Posts can become outdated quickly

Summary:
Knowing the difference between posts and pages helps create a well-organized site that's both dynamic and structured.

WordPress Gutenberg Editor

Definition:
The Gutenberg editor (also known as the block editor) allows users to build content using drag-and-drop blocks instead of one large text area.

Key Components:

1. Content blocks (paragraph, image, list, etc.)
2. Reusable blocks
3. Drag-and-drop design
4. Full-site editing (FSE) support

Benefits:

1. Beginner-friendly
2. Flexible layout control
3. Faster content formatting
4. Visual preview of content

Examples:
You can create a landing page with headers, images, and buttons without using HTML.

Challenges:

1. Not all plugins/themes support it
2. Some users prefer the Classic Editor
3. Can be buggy with third-party blocks
4. Learning curve for advanced formatting

Summary:
Gutenberg gives users more design freedom with a visual interface, making it easier to create stylish, engaging content.

WordPress Permalinks

Definition:
Permalinks are the permanent URLs of your posts and pages, controlling how site links are displayed.

Key Components:

5. Structure settings (e.g., /blog/%postname%/)
6. SEO-friendly URLs
7. .htaccess file (for Apache servers)
8. Slug customization

Benefits:

9. Improves search engine visibility
10. Clean and readable URLs
11. Easy sharing of links
12. Helps with site structure

Examples:
`yoursite.com/2025/ai-dictionary-release/` is cleaner than `yoursite.com/?p=123`.

Challenges:

13. Changing permalink structure can break links
14. Requires technical knowledge to fix redirection issues
15. Poor structure can confuse users and search engines

Summary:
Permalinks are essential for SEO and user navigation. Set them up properly at the start to avoid future headaches.

WordPress Media Library

Definition:
The Media Library is a central place to upload, store, and manage all image, video, audio, and document files on your site.

Key Components:

16. Upload button
17. File sorting and filtering
18. Image editing tools
19. Attachment details

Benefits:

20. Keeps media organized
21. Allows reuse of files
22. Reduces need for external hosting
23. Simple interface

Examples:
You upload your logo once and insert it into multiple pages and posts.

Challenges:

24. Limited file types unless adjusted
25. Large files can slow down site speed
26. Needs regular cleanup to prevent clutter

Summary:
The Media Library makes it easy to manage your visuals and documents, creating a seamless content workflow.

WordPress Menu Navigation

Definition:
Menus in WordPress are custom navigation links that organize pages, posts, categories, and more for users.

Key Components:

27. Menu creation panel
28. Drag-and-drop menu builder
29. Dropdown and nested items
30. Theme location assignment

Benefits:

31. Enhances user experience
32. Highlights important content
33. Flexible menu locations
34. Supports custom links

Examples:
A top menu with "Home," "About," "Shop," and "Contact" links helps visitors explore your site.

Challenges:

35. Poorly structured menus confuse users
36. Limited design if theme doesn't support customization
37. Complex sites may need mega menus

Summary:
Menus guide your visitors. Keeping them simple and well-structured makes it easy for users to navigate your content.

WordPress User Roles

Definition:
WordPress user roles define the permissions and capabilities of each registered user on your site.

Key Components:

1. Administrator, Editor, Author, Contributor, Subscriber
2. Role assignment
3. Plugin-based role customization
4. Capability control

Benefits:

1. Maintains security by limiting access
2. Organizes content responsibilities
3. Supports collaboration
4. Can be extended with plugins

Examples:
An author can write posts, but only an editor or admin can publish them.

Challenges:

1. Misassigned roles can lead to security risks
2. May need advanced setup for custom permissions
3. Role limitations may require extra plugins

Summary:
User roles are key to managing access and teamwork. Set them carefully to maintain control and protect your site.

WordPress Widget

Definition:
A widget is a small content block that can be added to sidebars, footers, and other widget-ready areas.

Key Components:

4. Drag-and-drop interface
5. Widget-ready theme areas
6. Content elements like menus, text, or calendars
7. Plugin-generated widgets

Benefits:

8. Easy to add features like recent posts or contact forms
9. Enhances site interactivity
10. No coding needed

11. Custom placement options

Examples:
A footer widget might include your business contact info. A sidebar widget could show your latest blog posts.

Challenges:

12. Not all themes support widget areas
13. Can become cluttered if overused
14. Limited layout control
15. Widgets don't always display consistently on mobile

Summary:
Widgets offer quick customization for non-tech users, helping enhance user engagement and design without coding knowledge.

Themes

Definition
Themes control the design, layout, and appearance of a WordPress website.

Key Components

1. Theme Files (style.css, index.php, etc.)
2. Theme Customizer
3. Child themes (for safe customization)

Benefits

1. Pre-designed layouts save time
2. Responsive design support
3. Easy to switch and preview different looks

Examples
A wedding planner may choose an elegant theme with a gallery homepage, while a news site may use a magazine-style theme.

Applications
Used in personal blogs, eCommerce stores, corporate websites, and portfolios to create a unique visual identity.

Challenges

1. Poorly coded themes can slow down the site
2. Switching themes may break layout consistency

Summary
Themes are the face of your website, offering design freedom and brand expression with minimal technical effort.

Widgets

Definition
Widgets are small content blocks that can be added to specific areas of a site, like sidebars and footers.

Key Components

1. Widget-ready areas
2. Drag-and-drop interface
3. Custom HTML widgets
4. Third-party widget plugins

Benefits

1. Easy to display calendars, recent posts, menus
2. Enhances user navigation and experience
3. No coding required

Examples
Adding a search bar widget to the sidebar or a recent posts widget in the footer.

Applications
Widely used in blogs, portfolios, and magazine websites to improve site layout and functionality.

Challenges

1. Limited design options without custom coding
2. Some themes restrict widget placement

Summary
Widgets add interactive and dynamic content to WordPress sites, enabling customization without touching code.

WP-Cron

Definition
WP-Cron is WordPress's built-in scheduling system for automating tasks such as publishing posts, checking for updates, or running backups.

Key Components

1. Hook-based system
2. Scheduled tasks and event handlers
3. Triggered on site activity (not real-time)

Benefits

1. Automates repetitive actions
2. No need for external scheduling services
3. Essential for time-based functions

Applications
Used to schedule content, perform cleanups, send emails, and manage backups.

Industry
Publishing, ecommerce, membership platforms

Challenges

1. Can fail on low-traffic sites
2. Not a true cron job (depends on page load)

Summary
WP-Cron provides automation for key WordPress tasks, helping maintain site operations without manual intervention.

LETTER X

XML (Extensible Markup Language)
A markup language used to structure and store data in a format that is both human-readable and machine-readable. In WordPress, XML is commonly used for importing and exporting site content through tools like the WordPress Importer.

XML Sitemap
A file that lists all important pages on a website to help search engines like Google index the site more effectively. WordPress SEO plugins such as Yoast SEO or Rank Math automatically generate XML sitemaps.

XHTML (Extensible Hypertext Markup Language)
A web coding standard that combines the rules of XML with the flexibility of HTML. While it is less common today, some older WordPress themes and websites still use XHTML for compatibility.

X-Frame-Options
A security HTTP header that helps prevent clickjacking attacks by controlling whether a website can be displayed inside an iframe on another site. WordPress security plugins can help manage these headers.

XSS (Cross-Site Scripting)
A type of security vulnerability where malicious scripts are injected into web pages. In WordPress, protecting against XSS involves keeping plugins, themes, and core files updated, and using security measures like input sanitization.

LETTER Y

YAML (YAML Ain't Markup Language)
A human-readable data serialization format often used for configuration files. In WordPress, YAML may be used in theme or plugin development for storing settings in a structured format.

Yearly Recurring Payments
A subscription payment model where customers are billed once per year for access to a product or service, often used in WordPress membership or e-commerce sites.

YouTube Embed
The process of inserting a YouTube video directly into a WordPress page or post using the video's embed code or the built-in WordPress embed feature.

Yoast SEO
A popular WordPress plugin for optimizing websites for search engines. It provides tools for managing meta tags, sitemaps, readability, and keyword targeting.

Yield Optimization
A strategy for maximizing revenue from website traffic, often applied in WordPress sites with ads or affiliate links by adjusting placements, layouts, and content for better conversion rates.

LETTER Z

Zero Downtime Deployment
A method of updating or changing a WordPress website without causing any service interruption, ensuring users experience no outages during maintenance or upgrades.

Zip File
A compressed file format (.zip) often used to package and upload WordPress themes, plugins, or backups for easy transfer and installation.

Zone File
A file stored on a DNS server containing information about a domain's IP address and related records, which is essential when pointing your domain to your WordPress hosting.

Zoom Integration
The process of connecting Zoom with a WordPress site, often used for webinars, online classes, or virtual meetings via plugins or APIs.

Z-Index
A CSS property that controls the stacking order of elements on a WordPress site, useful for ensuring menus, popups, or banners display correctly above other content.

WordPress Dashboard
Dashboard
Home
Updates
Add New
Categories
Tags
Posts
All Posts
Add New
Categories
Media
Pages
All Pages
View Comments
Comments
Vew Comments
Themes
Customize
Widgets
Menus
Theme File Editor
Appearance
Themes
Add New
Plugins
Plugin File Editor
Users
All Users
Add New
Profile
General
Writing
Reading
Discussion
Media
Permalinks
Privacy
Tools
Settings
General
Writing
Reading
WPForms
LiteSpeed Cache
3D FlipBook
3D FlipBook
Solid Security
Solid Security

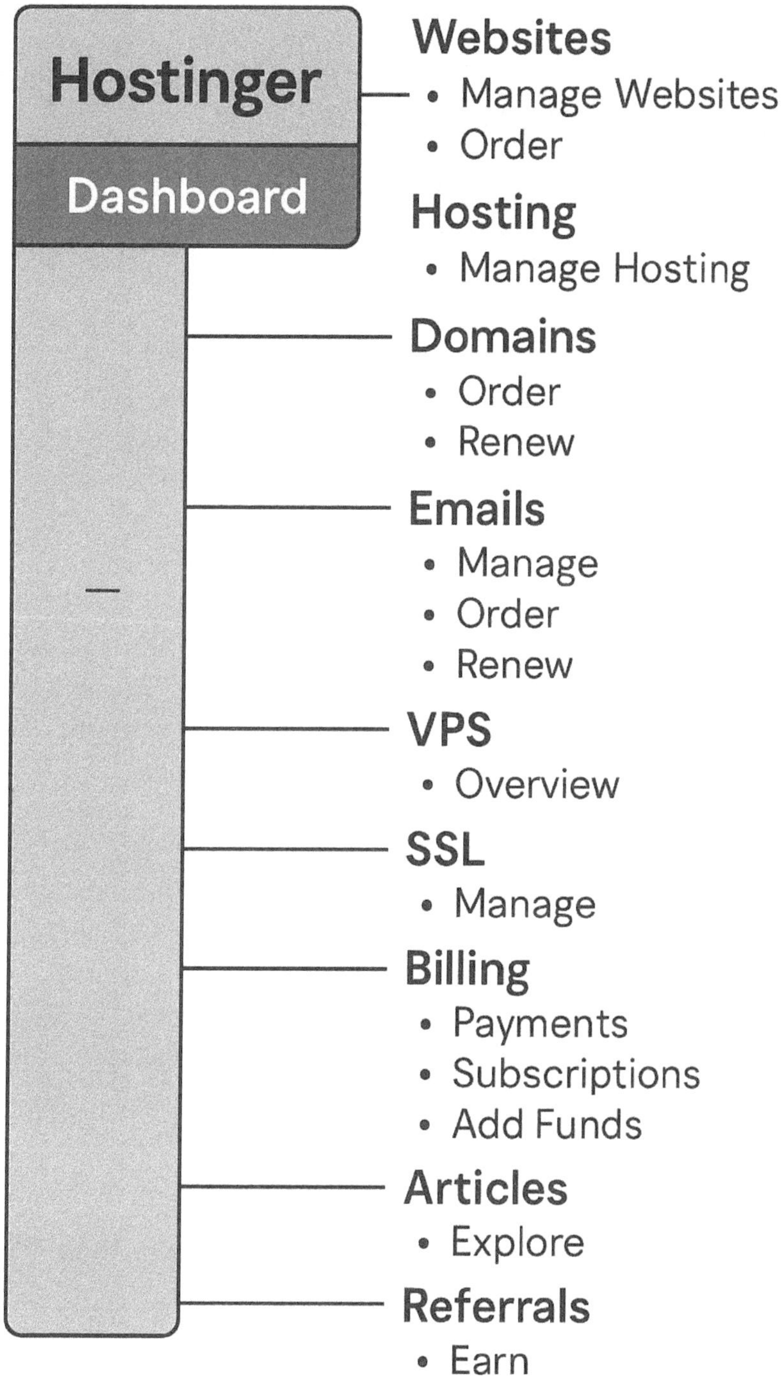

Payment Gateway Companies for Your E-commerce Store

Choosing the right payment gateway is crucial for your online store, ensuring secure, seamless transactions for customers buying products like my *IntelliGloss: The Definitive AI Dictionary* on youraistudybuddy.com. As a non-technical entrepreneur, I struggled with WordPress's hidden menus and Hostinger's unclear labels, so this list includes beginner-friendly payment gateways that integrate with WooCommerce and WordPress, support subscriptions, and work globally. Use these to process payments for your store, whether you're selling books, courses, or digital content.

Payment Gateway	Description	Example	See Also
Stripe	A developer-friendly gateway offering seamless WooCommerce integration, supporting 135+ currencies and recurring billing for subscriptions. It's known for its security (PCI compliance) and ease of use.	I use Stripe on youraistudybuddy.com to process *IntelliGloss* subscription payments, with my AI Study Buddy chatbot guiding users to checkout in 20+ countries.	*Stripe: Connecting to Your Website*
PayPal	A globally trusted gateway with over 430 million users, offering easy setup, no monthly fees, and support for credit cards, digital wallets, and PayPal Pay Later.	My PayPal integration on youraistudybuddy.com lets customers pay for *IntelliGloss* with one click, boosting trust for international buyers.	*WordPress: Setting Up Payments*
Square	Ideal for businesses with physical and online stores, Square supports WooCommerce, recurring payments, and digital wallets like Apple Pay, with straightforward fees.	I could use Square for *IntelliGloss* sales at pop-up events, syncing online and in-person transactions on youraistudybuddy.com.	*WooCommerce: Managing Transactions*
Authorize.net	A reliable gateway with strong security and WooCommerce integration, supporting multiple currencies and recurring billing for subscriptions.	On youraistudybuddy.com, Authorize.net could securely process *IntelliGloss* payments, with fraud detection protecting my global customers.	*WordPress: Setting Up Payments*
Amazon Pay	Allows customers to use their Amazon accounts for quick checkouts, supporting 170+ countries with robust security and subscription options.	Amazon Pay on youraistudybuddy.com lets *IntelliGloss* buyers check out using their Amazon credentials, streamlining purchases.	*WooCommerce: Managing Transactions*
Braintree	A PayPal-owned gateway offering free recurring billing, WooCommerce integration, and support for Apple Pay, Google Pay, and 130+ currencies.	Braintree's vault feature could store *IntelliGloss* subscriber details on youraistudybuddy.com for faster checkouts, reducing cart abandonment.	*Stripe: Connecting to Your Website*
Worldpay	A scalable gateway for high-volume stores, supporting 300+ payment methods and 146 countries, with advanced fraud protection.	Worldpay could handle *IntelliGloss* payments for my global audience on youraistudybuddy.com, ensuring secure international transactions.	*WooCommerce: Managing Transactions*
Alipay	A leading gateway for Chinese consumers, supporting WooCommerce	I could add Alipay to youraistudybuddy.com to sell *IntelliGloss* in China, tapping	*WordPress: Setting Up Payments*

	and cross-border payments, ideal for targeting Asia.	into 1.3 billion users via my AI Study Buddy's guidance.	
Payoneer	A global payment platform supporting 200+ countries and 150 currencies, ideal for freelancers and small businesses with WooCommerce integration.	Payoneer could process *IntelliGloss* payments for my freelance contributors on youraistudybuddy.com, simplifying international payouts.	*Stripe: Connecting to Your Website*
Razorpay	A popular Indian gateway supporting WooCommerce, with 150+ payment methods (e.g., UPI, cards) and competitive fees for global	Razorpay could enable *IntelliGloss* sales in India on youraistudybuddy.com, with my AI Study Buddy promoting local payment options.	*WooCommerce: Managing Transactions*

Love this direction. Let's stitch your WordPress book together with **clear internal links** so readers can hop from your "numbers & structure" chapters into the digital basics (bit/byte/CPU/GPU/TensorFlow) without getting lost.

Site map & slugs

Use these **page titles** and **slugs**; paste the one-line blurbs under each page's intro for clarity.

1. **Numbers & Structure (Start Here)** — `/numbers-structure`
   - *How our everyday counting (single → double → triple digits) builds place value and "6/7/8-figure" thinking.*
2. **Binary for Humans** — `/binary-basics`
   - *Same structure as decimal, but the places are 1, 2, 4, 8, 16… instead of 1, 10, 100, 1,000.*
3. **The Bit Family (Bit → Byte → KB → … → YB)** — `/bit-byte-family`
   - *A friendly ladder from a single bit to yottabytes, with real-world examples of text, photos, songs, and movies.*
4. **What Is a Chip?** — `/what-is-a-chip`
   - *From raw silicon to a finished microchip: the "muscle" that runs software and AI.*
5. **CPU vs GPU (Who Does What?)** — `/cpu-vs-gpu`
   - *CPU = versatile planner; GPU = thousands of helpers in parallel. Why AI loves GPUs.*
6. **Memory 101: DRAM vs VRAM vs HBM** — `/memory-dram-vram-hbm`
   - *Where data lives while it's working: system RAM vs ultra-fast GPU memory.*
7. **CUDA in Plain English** — `/cuda-explained`
   - *NVIDIA's bridge that lets software use the GPU for non-graphics math (AI, science, simulation).*
8. **TensorFlow (The AI Toolkit)** — `/tensorflow-basics`
   - *Open-source tools to build, train, and deploy machine-learning models.*
9. **AI's Energy Appetite** — `/ai-energy`
   - *Watts, kilowatt-hours, cooling, and why data centers care about efficiency.*
10. **Glossary (Quick Look-ups)** — `/glossary`

- *Fast definitions with links back into the right chapter.*

Cross-link plan (put these at the end of each page)

Add a **"Keep Learning"** box with 3–4 contextual links.

On `/numbers-structure`:

- Next: **Binary for Humans** → `/binary-basics`
- See: **The Bit Family** → `/bit-byte-family`
- Jump to: **CPU vs GPU** → `/cpu-vs-gpu`

On `/binary-basics`:

- Build intuition: **The Bit Family** → `/bit-byte-family`
- Hardware bridge: **What Is a Chip?** → `/what-is-a-chip`
- AI angle: **CPU vs GPU** → `/cpu-vs-gpu`

On `/bit-byte-family`:

- Where bytes get used: **Memory 101** → `/memory-dram-vram-hbm`
- AI software: **TensorFlow** → `/tensorflow-basics`
- GPU software: **CUDA** → `/cuda-explained`

On `/cpu-vs-gpu`:

- Memory matters: **DRAM vs VRAM vs HBM** → `/memory-dram-vram-hbm`
- Use the GPU: **CUDA in Plain English** → `/cuda-explained`
- Try frameworks: **TensorFlow** → `/tensorflow-basics`

On `/cuda-explained`:

- Hardware partner: **CPU vs GPU** → `/cpu-vs-gpu`
- Build models: **TensorFlow** → `/tensorflow-basics`
- Data size context: **Bit Family** → `/bit-byte-family`

On `/tensorflow-basics`:

- Speed it up: **CUDA** → `/cuda-explained`
- Hardware basics: **CPU vs GPU** → `/cpu-vs-gpu`
- Data scale: **Bit Family** → `/bit-byte-family`

On `/ai-energy`:

- Why energy rises: **CPU vs GPU** → `/cpu-vs-gpu`
- Memory heat: **DRAM vs VRAM vs HBM** → `/memory-dram-vram-hbm`

Ready-to-paste WordPress blocks

"Keep Learning" callout (Gutenberg)

- Add a **Group** block → **Heading** "Keep Learning" → List with links:

```
Keep Learning
• Binary for Humans → /binary-basics
• The Bit Family → /bit-byte-family
• CPU vs GPU → /cpu-vs-gpu
```

Inline "see also" callouts

Drop this inside paragraphs where terms appear:

- *See also:* The Bit Family · CPU vs GPU · CUDA

Reusable block (make once, reuse everywhere)

Create a reusable block named **"Next Steps (AI Basics)"**:

```
Next Steps (AI Basics)
→ Binary for Humans (/binary-basics)
→ Bit → Byte → KB → … (/bit-byte-family)
→ CPU vs GPU (/cpu-vs-gpu)
→ CUDA explained (/cuda-explained)
→ TensorFlow basics (/tensorflow-basics)
```

Breadcrumbs (helps structure)

Install a breadcrumbs plugin (or theme feature) and use this pattern:

- **Home › Numbers & Structure › Binary for Humans › The Bit Family**
- **Home › Hardware › CPU vs GPU › Memory 101**
- **Home › AI Software › CUDA › TensorFlow**

Suggested categories:

- **Foundations** (numbers, binary, bit family)
- **Hardware** (chip, CPU/GPU, memory)
- **AI Software** (CUDA, TensorFlow)
- **Big Picture** (energy)

Index AI and Tokens

KO E 'OTUA MO TONGA KO HOKU TOFI'A

www.ingramcontent.com/pod-product-compliance
Ingram Content Group UK Ltd.
Pitfield, Milton Keynes, MK11 3LW, UK
UKHW062012041026
15591UKWH00017B/1296

9 798999 196316